To

Raymond Cooper -

June 25 - 1989

[illegible]

TILLING THE GOOD EARTH

by

Dan Logan

Library of Congress Catalog Card Number: 89-083521
ISBN: 0-944419-08-9

First Edition
Printed 1989

Tilling the Good Earth

By: Dan Logan

For information contact:

The Everett Companies
Publishing Division
P. O. Box 5376
Bossier City, LA 71171-5376

Phone: (318) 742-6240
LA WATS (800) 826-6512 NATL WATS (800) 423-7033

Published in the United States of America
by: The Everett Companies
Publishing Division
813 Whittington Street
Bossier City, Louisiana 71112

① 2 3 4 5

ACKNOWLEDGMENT

With grateful appreciation to
Dr. Frank Lower
for his help in preparing
this manuscript

Table of Contents

FOREWORD

A man past 80 has the unique privilege of speaking the truth as he sees it. It will not be long before he will be called on to speak before his Maker and give an account of his actions while he was allowed to roam and walk to and fro here on this planet. He will have to explain what he did with the talents that he had while here.

Now there are a few people who don't believe they will have to give this account but as for me I do not want to take any chances.

When old Gabriel blows his horn and announces that everyone come forward from where we are and stand at attention and wait until our turn comes to speak up and explain what we did with the talents given us, I just naturally want something to say that might help the present cause.

From the beginning of time, famine has dogged the footsteps of man and sometimes there was enough to eat and sometimes there wasn't. Most of the time it was just a get by. This existed until just recently when universal education was established in this country and other sources of power were discovered or devised such as steam, electric, gasoline, diesel, and atomic power. From this man has learned mass production, rapid transportation and ways to preserve food. All this together has made food available to people of the whole world. But more is still needed. I am trying to improve farming.

There is not a single person who won't benefit by better production for we all depend on food.

Better education is needed in our colleges so farmers can do a better job.

I have farmed for 60 years. The information in this books comes from my lifetime of observation and also from those people I discussed farming with 50 or 60 years ago. They had had a lifetime of experience which makes the book cover more than 100 years of observations. I learned from such men as John Glassell, a big planter, who was old in 1930 and had accumulated wisdom in farming all his life. From John Sentell, who was a great farmer and man, also a large land owner and about mt. Glassell's age. From John Caldwell, an ex-slave, and large land owner. He was a big boy and drove the carriage for Mr. Masters' wife during the Civil War. Frank Marshall (a negro who came from South Carolina to this country when he was a lad). Frank possibly was the best farmer I ever knew. His opinion was always correct. Everyone else recognized it, too, and he was a share cropper and a fine upright man.

From Mr. Birdwell, a professional plantation manager, whom I never knew, but all the older, good farmers did what Mr. Birdwell said to and I tried to figure out why.

The first twenty-five years of my farming life were more conducive to learning than the last twenty-five years or so for the reason that we dealt in smaller crops per person and more people, more involved, who had learned by experience. People must be encouraged to learn from others than just learning the hard way by trial and error.

Fifty years ago a farmer was ashamed to make a sorry crop. When his neighbor made a good crop and he made a sorry one his feelings and pride were hurt. Today this does not bother him. If his crop is bad, he just lays it on the government for not giving him the right information. If the crop is not enough to pay expenses, he just explains that the government must raise the loan rate. The weather did not cooperate at all. Local dry conditions just came at the wrong time.

It never occurs to him that he has any responsibility for the production on his land under his control. It is always something beyond his control.

They just go broke for lack of information that will make them free and prosperous. If the farmer would look at conditions himself, he could go forward.

He must observe and see the results himself to see if they are correct. If he does not observe and see his own mistakes and have faith enough to follow his decision, it is probably better that he find another way to make a living. It seems today that the only requirement for a farmer to be successful from the beginning of his farming career is for him to want to be successful enough to follow the proven methods. To be a successful airplane pilot and not wreck the plane the student takes instructions from a pilot who has proven that he can fly and listens to him until he learns how to take the plane off and bring it back to the field successfully. His instructor stays with him until he can do this himself safely, before he lets him solo.

Today there is so much good equipment. There are so many good and efficient chemicals there is no need for a failure to be made if the farmer knows how to make the right decision and makes it at the right time. Very rarely is the weather the cause of the failure even though most all failures are blamed on the weather and no one can dispute this claim. If the truth be known, very few failures are caused by the weather, but if anyone tells a farmer he failed because he did not know or he did not listen to advice from old John or Jack or James, his neighbor who made a good crop when he failed, you then have made an enemy and this can never be proved so just don't say it. The decisions a farmer makes is what makes him a profit or a failure. It is not the amount of money he spends, but the decisions he makes.

It is almost impossible to be a successful and prosperous farmer without common sense. This priceless ingredient is to learn what do do and then apply this learning on time. At this point the men are separated from the boys.

In agriculture there have been men in all ages who have learned to do a better job than those around him, but this know-how has never been written down and documented so others can take up where he has left off. When the persons who have learned these secrets of Nature die, then this know-how is lost. This should not be. This knowledge should be recorded and saved. It seems logical that the necessary thing can be learned and taught, how to transfer experience from one to another, from one who has experience to the one who wants to learn what the experienced person knows.

Information has increased so much in the last 100 years. There is no way for anyone to know all there is to know. A person to know one thing well must have the courage to be ignorant of thousands of other things regardless of how interested that person is in them.

The same thing applies to work. In order to do one thing well a person must single out a specialty. This common job must done in an uncommon way.

Nature teaches men through their mistakes and uses their distress to open their eyes on how to do a better job.

TILLING THE GOOD EARTH

In this book I have recorded secrets of nature that I have learned from years of farming. May you learn from my experiences and prosper in your farming.

Dan Logan

The author in his office on the farm. He has never had a dull day in all his farming life. There is always interesting things to do and plenty to challenge you. He is at present teaching a class on practical farming at La. Tech University and was made a professor after he was 80 years old.

About the Author

For sixty years Dan P. Logan has been farming the same Red River bottom land at Gilliam, Louisiana. The lessons he has learned from trial and error he is sharing with farmers everywhere.

He was born July 4, 1907 in Benton a small northwest Louisiana town. His father died when he was seven and his mother died when he was 12. After his mother's death he lived around with different friends and relatives.

Although he did not graduate from high school at 18 he enrolled in Tulane in New Orleans. He did not have the foundation to handle the school work and had to drop out. He tried to go to Centenary College in Shreveport while working in a filling station to pay his way. The work was too demanding and he had to quit school. He still wanted an education but he had to learn on his own.

He asked for a job on his uncle's plantaton at Gilliam, La. He lived with his uncle and worked there for 3 years for $1 a day. Dan P. Logan took over more duties as he got older. He became overseer of the place. Later he bought the property and reared his family in that same plantation home.

Although Dan P. Logan started his career with disadvantages-he had lost his parents at an early age and he had had little formal education – he has accomplished more than any man in his area. He has bought and sold cotton since 1930 and founded the Logan Cotton Company. He worked with the LSU Experimental Stations on his farm and grew seed corn. He started the Logan Seed Company. The farm land needed fertilizer. He started the Farmer's Ammonia Company. Growers had to take their picked cotton a long distance to have it ginned. He built the Gilliam Gin Company and Valley Delinting Company. There were other farming related enterprises.

The richest land could not be farmed because it was so low that water stood on it. In the early forties Dan P. Logan got a patent on dirt moving machine. This enabled farmers to drain their land. The hills, which were eroding from cultivation, were allowed to grow up in trees by their owners and more bottom land was cleared for farming.

In 1947 Dan P. Logan started the first farm consulting service for insect control. He bought a plane and learned to fly and built an airfield on the farm. He employed bug checkers and served all the farmers from Shreveport to Idabel, Oklahoma, spraying for insects.

He started with no money at all and became a successful plantation owner with many business connections. He accomplished this by hard work and filling needs to make farming more efficient. He still found time to spend in his fields observing and recording information about his crops.

For 60 years the top men in Agriculture have been his friends. Many new ideas have been tried out on his farm. He pioneered in irrigation in North Louisiana, worked with the first cotton picking machine and tried out new ideas in weed control, follier feeding, petiole analysis and the use of sugar in feeding the cotton plant.

At age 81 he has a branch of Louisiana Tech stationed on his farm and is teaching a course at Louisiana Tech on Farming.

Dan P. Logan knows that is not enough to get a degree in Agriculture, that you learn from experience important lessons in farming that have not been previously printed. In *Tilling The Good Earth* he is sharing some of this valuable information.

LOUISIANA TECH UNIVERSITY

College of Life Sciences

The Department of Agricultural Sciences, Technology and Education takes pleasure in recognizing your contribution of professional expertise in the field of cotton production and for your services as adjunct professor in the instruction of agricultural students in a Special Problems Course on "Nature's Laws as They Apply to Agriculture."

Presented to

Dan P. Logan

in deep appreciation on this 5th day of December, nineteen hundred and eighty seven, at Ruston, Louisiana.

Hal B. Barker, Dean
College of Life Sciences

Daniel D. Reneau, President
Louisiana Tech University

Larry D. Allen, Head
Department of Agricultural Sciences, Technology and Education

COLLEGE OF LIFE SCIENCES DIPLOMA

A Few Friends Speak About Dan Logan's Ability As A Successful Cotton Farmer

105-F East 30th Street
Austin, Texas 78705
August 23, 1988

Mrs. F. C. Boston
5406 Fairfax
Shreveport, LA 71108

Dear Mrs. Boston:

Thank you for your letter of June 29. I am sorry I have been so long in responding. Hopefully it will not be too late.

I farmed at Natchitoches for eight years beginning in 1978, and early during that time Dan Logan became my dear friend. He and I were "traveling buddies", and would take off during breaks in farm work, and travel around the Cotton Belt. He and I have traveled all over Louisiana, Arkansas, and Mississippi together. And we have been to the Yuma Valley of Arizona, and the Imperial Valley and San Joaquin Valley of California. Everywhere we have been there is no farm home that Dan Logan is not welcome in.

I recall one routine trip Dan and I made to inspect the crops of northwest LA, and we stopped unannounced at the farm of former Governor John McKeithen at Columbia, since Dan had heard that the governor was trying something new he wanted to see.

While on the farm we encountered the former governor in his jeep, and he invited us to his home for lunch. Governor McKeithen fixed chicken and dumplings himself, and the three of us sat around the kitchen table and ate and talked cotton and politics. I was somewhat amazed by the scene, but it was routine for Dan.

Another sight which is vivid in my memory is of Dan on top of one of the cotton pickers. During harvest time Dan would come down to see how we were doing, and he always wanted to take pictures of the crop. To get the best possible picture of the crop during harvest this 75-year-old man would climb up on top of the cab of one of my International 782 cotton pickers while it was moving through the field. There he would lie spread-eagled on his stomach, peering over the front edge of the cab, and begin snapping pictures of the cotton below.

As representative of Dan's influence on me, I want to share with you the major part of a homily I preached this summer at Houston's M.D. Anderson (Cancer) Hospital. I was serving as one of the chaplains at the hospital this summer, and it was my responsibility to preach in the hospital's chapel on Sunday, June 19, 1988. The Dan in this homily is, of course, Dan Logan. There were 43 people in attendance at this service, and it is yet another anecdote that at the conclusion of the service one man came up to me and thanked me for the message and said, "I think I know the Dan you're talking about." He did, too.

I have included a copy of the major part of this homily. You may use any or all of it in the book, as you see fit. My position at the time was Chaplain Fellow, M.D. Anderson Hopital. I am now back in Austin, where I am Senior at Austin Presbyterian Theological Seminary.

If I can reduce what Dan taught me to a single statement, it is that for life to be meaningful it must be guided by a vision. Dan Logan has lived by a vision, a vision that served to invite us all to move beyond where we are, and into God's future.

I hope I have been of some help.

Sincerely

George A. Armstrong
Senior, Austin Presbyterian
Theological Seminary

The Freeman-Dunn Chapel
M.D. Anderson Hospital

Order of Morning Worship
June 19, 1988

Call to Worship		The Shema, Deut. 6:4-9
Hymn	"Come Thou Almight King"	#140
Responsive Reading		Psalm 130, #51
Prayer for Illumination		
Old Testament Lesson		Ezekiel 17:22-24
Gospel Lesson		Mark 4:26-34
Sermon:		
	Why Not Say, "To Hail With It"	
		Chaplain George Armstrong
Prayers of the People		
	"A Litany of Thanksgiving"	#120
Our Lord's Prayer		
Hymn	"Be Thou My Vision"	#225
Charge and Benediction		

M.D. Anderson Hospital **June 19, 1988**

WHY NOT SAY "TO HAIL WITH IT"?

(Mark 4:26-34)
(the main body of a homily
preached by George Armstrong)

In our text this morning Jesus talks about the kingdom with two parables: the familiar parable of the grain of mustard seed, and perhaps the less familiar parable of a person who sows seed and seems rather passively to watch what happens.

I am grateful to have this agricultural sounding text this morning, a text which is being preached in many Christian churches worldwide today. You see, for eight years I was a farmer, a cotton farmer in the Red River Valley of NW Louisiana.

So, both as a farmer and as a chaplain I studied the text. Particularly the first parable captured my attention. A farmer plants a crop – sort of just throws it out there. He doesn't know how it does what it does but he must have confidence in it or he wouldn't have planted it. And the plants do what they are intended to do: they produce the blade of leaf, then the ear or flower, and then the payoff – the fruit. That's what the plant lives for: to produce fruit, the bearer of seed.

And farmers seem to have neverending confidence that is going to take place – that the bountiful harvest will come in. Farmers are the eternal optimists. I am reminded of the old timer who was asked about his best crops. He thought for a moment and replied. "In all my years of farming I've had two good crops: way back in 1953; and next year."

This parable caused me also to think of one of my own crops: the 1983 crop. We planted the cotton right on time – May 1 – and five days later we had a perfect stand: 2.7 plants per foot. (Sometimes I do silly things like count cotton plants.) With a good stand you are well on your way to a good harvest. Within 10 days the cotton was about this high and was putting on its first true leaf.

And then we had a storm. Not your little summer-time shower, but the kind we are prone to get in May when the air is unstable and the weather pattern is in transition. We got one of those storms. Rain: five inches in two hours, some people call it a "toad-strangler" or a "chunk floater". And hail – tremendous hail, hail that beat the devil out of that cotton. The leaves the hail didn't strip off it shreaded like cabbage.

I've seen pictures of farmers in their storm-ravaged or drought-stricken fields, but this was the first time I stood in my own. It was pitiful. Half the cotton was dead, and the other half might as well have been. It was laying over in the mud with its leaved stripped off by the hail and the soil washed from around its roots by the torrential rain.

Most of my neighbors started gearing up to replant their crops as soon as it was dry. Maybe that was the right thing. But I hadn't been farming very long and I didn't have much sense. So I call my 75-year-old farmer friend, Dan, from 90 miles up river, north of Shreveport, and he came down and looked at this pitiful mess with me. He didn't have the fancy college degrees, but he had the kind of education you get from 65 years in a cotton field, from watching the ways of nature, which seem to bless you one day and curse you the next, and from watching the old black man next door who was farming with a mule and a double-shovel long before Dan himself got started.

Well my friend Dan and I agreed that we would keep the crop. We would tend it best we could, and see what it would do. Some people might say that you don't have much invested in a crop just planted so you might as well start over. But that wasn't the basis for a decision. The basis for a decision was this: what is that crop's potential? That crop lay out there in the field, half of it dead, and the other half just as well as dead – what is the potential of that crop? Some people would laugh. Some people probably did laugh.

But my friend Dan – that old man knew what a lot of people didn't know. Oh, he saw the pitiful mess lying on top of the ground. But he also knew what was underneath. He knew the root system of that plant. A cotton plant ten days old already has a substantial root system, and a tap root that goes down deep. Deep inside, inside the soil, has been developed a root system that can sustain that cotton plant in good times and in bad. Nobody sees it. Some doubt if it's really there. But it is. The farmer didn't put it there. We might say Mother Nature did.

So we kept that crop. There wasn't much we could do for a couple of weeks. We just watched and waited. Some more of the cotton died of seedline disease, but most of it hung on precariously. After a few days we could get down on our knees and see the tiniest little green bud appear where a leaf used to be. A new leaf. It's as if the cotton plant is starting all over. But you see, it's not restarting over. This cotton plant is going to use its whole life experience, good and bad, to do what it's intended to do, to bear fruit. And it's even going to make up for lost time. Yes, on the surface it's mangled and scarred, but it's got a root system that goes deep. And a cotton plant is not judged by what it looks like laying in the mud, or the scars on its trunk which it will carry even to the day of harvest. A cotton plant is judged by the fruit it produces: the fluffy white bolls that will someday feel soft and warm on the skin of new born babies, and the seed in those bolls that will be planted for a new crop. The plant will die with the winter frost, but its fruit will live on in places we don't even know of.

So our management of that crop consisted of a combination of patient waiting and then doing what we could when we could. We could not do the impossible, but only respond faithfully to the ways of nature when they were beyond our control

What is the potential of a cotton plant? What is the intended purpose of God's creation? What is God's intention for us? These are not unrelated questions, neither are they easy to answer; for the Bible is full of metaphors from nature wherein the writers struggle to say something about God's ways. (Even Jesus used the language of nature to talk about God's Kingdom.) Often it is in the language of what is sometimes called the "divine reversal." Recall the language we read in Ezekiel. God says:

> *all the trees of the field shall know that I the Lord bring low the high tree, and make high the low tree, dry up the green tree, and make the dry tree flourish. (17:24)*

Translating that language into human terms is not easy, for God's ways often are not easily discernible, or physically visible. But the poetry of nature serves as an invitation for us to remain open, open to the God who is Lord even of the storm-ravaged cotton crop.

Oh, yes, that cotton crop. We harvested it in October. Of course, when we harvest it's not just a plant here or there that we harvest, but all the plants which together make the crop: all these plants which had been mangled and scarred way back in May. The crop made 1800 lbs. per net acre, two and one-half times the county average, and triple the state and national average. Thanks be to God!

Amen

28 June, 1988

Lengyel's
Agricultural Service
2417 East Indian School Road
P.O. Box 10441
Phoenix, Arizona 85064

Mrs. F. Boston
5406 Fairfax Street
Shreveport, LA 71108

Dear Mrs. Boston:

Your request to write a few words about Mr. Logan leaves me kind of stifled; I have known Mr. Logan since 1968 and because of that time, writing a book would be more apropos than just a few words; but at any rate, here goes nothing, and/or everything.

I got introduced to Mr. Logan with a phone call from him; naturally, he was calling about cotton in Louisiana. He had heard about my service (cotton petiole nutrient monitering) and he wanted to become acquainted with the service.

I had started the cotton petiole service with farmers and at that time it was the only laboratory in the United States offering that service to growers. Mr. Logan told me over the phone that what he was doing was wrong because the cotton grew up to seven feet tall and would only yield about a half a bale of cotton. Despite the tall cotton and the poor yield, the university recommendations were still 100# anhydrous ammonia preplant and another 100# around July 1.

Mr. Logan asked if he could come out to Phoenix to see what I was doing and I naturally said that he was welcome. I'll never forget the first time I saw Mr. Logan; he was getting off the plane...with a cane and obviously hobbling. My first question to him: Mr. Logan, how are you going to jump the irrigation ditches? Mr. Logan chuckled and said "I'll make it somehow."

And Mr. Logan did make those ditches and that I think typifies his character and attitude. He was always willing to try something new to grow a better cotton plant so the plant would yield more lint cotton.

Through soil and plant analysis, the nutrient fertilization was completely changed for Mr. Logan. He didn't apply any nitrogen for two years and his yield doubled to around 800-1,000 pounds. He was the first farmer in the Delta area to start foliar application of LB Urea to stimulate growth and to size up bolls and he was also the first farmer in the entire mid-south and south to use sugar to control vegetative growth of the plant and to induce more flowering and fruiting.

Otis Chapman, cotton grower and lay-pastor from Scott, Arkansas, illustrated Mr. Logan's attitude and devotion to the land with this story.

A couple years after I started working with Mr. Logan, I had a phone call from Otis Chapman and he told me this story. He had preached at a church in Louisiana on this particular Sunday and Otis wanted to look at cotton fields on the way home to Scott, Arkansas. Otis was interested (like Mr. Logan) in getting better yields.

Incidentally, Otis called me on the phone to relate this story to me of his trip back to Scott.

Otis told me that he had been traveling, looking at cotton fields but never stopping to look at any fields because none of the fields looked better than his own cotton fields back home. But, as Otis continued with his story, he said that when he came to a field being irrigated north of Shreveport, he stopped his car because it was the best cotton field he had seen all afternoon. A colored boy was irrigating the field and Otis asked him whose cotton it was. The colored boy's reply was somewhat unusual: "this here cotton belongs to Mr. Logan, but you don't want to talk to him. Mr. Logan is crazy, he uses sugar on his cotton." Without hesitation, Otis told the boy that if Mr. Logan was crazy, he wanted to be crazy too, just like Mr. Logan.

So Otis and Mr. Logan met, became good friends; and all because once again Mr. Logan was trying something new and improving the land. Otis naturally became my customer and through the intervention of Mr. Logan, his cotton yields also increased, by almost 100%.

Mr. Logan wasn't satisfied in just improving his own cotton yields; he personally set up meetings with farmers in the entire Delta areas and asked me to talk to all these growers. If anyone can be selected that promoted and helped establish cotton petiole monitering in the Delta, it would be Mr. Logan. He promoted and helped establish the present day Arkansas testing as well as the current proliferation of most of the laboratories that now offer this service. This is an indirect contibution because he himself didn't establish the laboratories but the laboratory personnel eventually saw the need by what Mr. Logan was doing along with all the other farmers that eventually accepted the service as part of their management of their cotton fields. Today, of course, it is an accepted part of most managements.

Because of Mr. Logan, growing cotton was changed in the Delta and yields increased. He awoke the desire for more research by universities. He carried on his own research on his own farm to show how yields can be improved. Without question, cotton and

Mr. Logan have become synonymous and his reflection with the plant is well established. In the future, when cotton production is discussed in the Delta, they will say of Mr. Logan, "he passed this way and made a better world for us all by his devotion to the land."

Sincerely

Albin D. Lengyel

OFFICE OF THE
CHANCELLOR

MAILING ADDRESS:
Post Office Box 25203
Baton Rouge, LA 70894-5203

July 5, 1988

OFFICE:
LSU Agricultural Center Bldg.
504 388-4161

Mrs. F. C. Boston
5406 Fairfax
Shreveport, LA 71108

Dear Mrs. Boston:

Mr. Dan P. Logan, farmer and businessman, Gilliam, Louisiana, is a good farmer, and a good friend of mine. He has an inquisitive mind and is continually searching for innovations to improve agronomic practices in cotton production. I have often said that if Dan had decided to pursue agricultural research ("instead of making money"), he would probably have been one of the most successful researchers in our state and nation.

My observation is based on a long personal acquaintance with Dan, dating back to the early 1970's. Dan invited special guests to tour his cotton production and to share the noon meal together. I visited him at other times during the cotton harvesting season as well.

Dan's favorite project was, and still is, the proper nutrition, or "feeding" of the cotton plant. This work involves the cotton petiole analysis to determine the nitrogen requirements for optimizing the fruiting, or boll formation, of the cotton plant and to control plant height within desired agronomic limits. As a result of his enthusiasm, I visited his farm many times to observe the results of his work. I also traveled with him to Arizona in the mid 1970's to visit his consultant and also see irrigated cotton fertilized according to the petiole analysis method recommendations.

A "cotton-wide" experiment in the use of petiole analysis was initiated by the Louisiana Agricultural Experiment Station in response to his hypothesis that the yield could be significantly improved by "feeding" the plant throughout the growing season. The accepted nitrogen fertilizer practice was, and still is, to apply part of the nitrogen at planting and part later as a side-dressing. The results of these tests were published as an Experiment Station bulletin and while noting the usefulness of petiole analysis as a measurement tool for nitrogen fertilization, it also demonstrated wide variability with this method because of the lack of soil moisture control.

Mr. Logan also did many other innovative things in cotton production. For example, I was impressed by his work with cotton irrigation and the efficiencies he obtained. He is a person who truly seeks knowledge, one who keeps records of each trial, and one who shares his knowledge with others. Yes, he is a good farmer, but a farmer who is not satisfied with the current research technology available to him and the cotton industry. Last year, 1987, was a good cotton year for Louisiana, with the highest ever average yields per acre. But, the yields were still not acceptable to Dan Logan. He knows we can do better. And, somehow I believe him! I look forward to seeing his dreams fulfilled.

Sincerely

H. Rouse Caffey
Chancellor

HRC:kps

Ray Young
Commercial Entomologist
Ag. Consultant
Farm
Wisner, Louisiana

Ray Young is one of the outstanding and oldest agriculture consultants in the nation. He has over 40 years experience and is a fairly large cotton planter himself. He is a consultant to others, managing over 30,000 acres. Ray has this to say about Dan Logan:

"Mr. Dan has done much for the Agriculture industry, especially cotton. He has inspired many with his dreams and visions of what could and should be done. I am grateful for the inspiration he has been to me which started in 1947 and continues even today."

INTRODUCTION

In India there is evidence that cotton was known and spun and woven over 4,000 years ago. It's safe to say it was cultivated in India 2,000 to 3,000 years ago. Information was scant. Each nation was isolated from the other. No one knows when cotton got started as a trader fiber. Marco Polo never even mentions it.

As late as the 15th and 16th century little was known about cotton in Europe. The seed was considered good for many ailments and a good remedy for poisoning. The oil of the seed would take away freckles, people believed.

The mode of growth of cotton led to several erroneous beliefs. Some people believed that cotton was the wool of little lambs that grew attached to the branches of a tree (see picture on page 36). John Mandeville, a man of great learning from Herefordshire, is said to have discovered this vegetable lamb. Mandeville had left his home in 1322 and spent 34 years traveling beyond Cathay and India to a kingdom called Caldeya and "there groweth a manner of fruit as though it were gourds, and when they be ripe men cut them in two and men find a little beaste in flesh and in bone and in blood as though it were a little lamb with wool outside it. Men eat both the fruit and the beaste and that is a great marvel. Of that fruit I have eaten, although it were wonderful, but I know that God is marvelous in all his works."

Another account of cotton is that a seed put into the ground grew a plant resembling a lamb and attaining a height of 2½ feet. It had a head, eyes, ears and all other parts of the body of a newly born lamb. The wool was extremely soft and used in the manufacture of head coverings. The lamb was rooted to the ground by a stem from the middle of the body (see picture) and ate the surrounding grass and plants. When the food supply was used up the lamb died.

We smile at the stories of the vegetable lambs but in 1988 we still have much to learn about the cotton plant.

1492 Columbus discovers America and found cotton growing here
1607 Colonists plant cotton in Virginia
1712 English parliament prohibits wearing cotton goods for it made the women look too great
1736 parliament gave up after 14 years and lets the women wear what they want
1769 Richard Arkwright patents method to weave cloth with machinery
1793 Eli Whitney invented the cotton gin. George Washington, president of the United States issued the patent to him

1845 Cotton went below 5¢ per pound. In 1931 it went below 5¢ again

1850 First patent issued for cotton-picking machine. It did not work

1861-65 Cotton production fell to 4% of its normal production

1892 The Boll Weevil moved into the Cotton Belt looking for a place to live

1916 Calcium Arsenate used to kill boll weevils

1925 New varieties developed for mechanical stripping

1926 11,299,000 bales exported

1927 John Rust made moisture spindle method of mechanical harvesting. Mechanical harvesting begun

1937 Biggest crop ever when 18,946,000 bales made on 33,500, 000 acres

1938 National Cotton Council organized

1946 First commercial use of organic insecticides

1955 Chemical weed control comes to farms

1963 Cotton Belt averaged better than 1 bale of cotton per acre

1973 Cotton price hit $1.00 per pound

What does the future hold for the cotton farmer?

This was the layman's perception of how cotton grew. They called it "vegetable wool" and believed the sheep looked like this.

Introduction

A NEW CONCEPT FOR THE USE OF COTTON

A fibrous substance similar to ordinary wool or flax

In Europe, about the fifteenth and sixteenth centuries, cotton was so little known that there were several superstitions believed regarding it. One was that its seed was useful as a cure for asthma, coughs, dysentery and wounds, and was a good remedy in case of poisoning. The oil of the cotton seed was recommended to take away spots or freckles, in fact, it was a cure-all and reminds us of a patent medicine circular of modern times.

Another curious superstition regarding it was its mode of growth: this was nothing more or less than that the **cotton wool** was really the wool of lambs that grew and lived attached to branches of trees. Of course, the only knowledge of fibrous substance possessed by our forefathers in those days was that of ordinary wool or of flax, and no doubt their first impresssion of any other fiber would be something that resembled the wool of the sheep or the hair of the goat. This belief was fostered or more probably established by one Sir John Mandeville, described as a man of learning and substance, of the town of St. Albans, in Herefordshire, who in the year 1322 left his native city, became a globe trotter, and did not return for thirty-four years.

In his report of his journey he states that he travelled through all the then known kingdoms of the world and among other things discovered this **vegetable lamb**. His account in his own words is: "Now shall I say you of countryes and isles that be beyond the countrye that I have spoken of. Passing beyond Cathay and India and Bachary is a kingdom that men call Caldeya, that is a fair countrye and there groweth a manner of fruit as though it were gourds, and when they be ripe men cut them into and find within a little beaste in flesh and bone and in blood as though it were a little lamb with wool outside it. Men eats both the fruit and the beaste and that is a great marvel. Of that fruit I have eaten, although it were wonderful, but I know that God is marvelous in all his works."

Another account is by Baron Von Herberstein, an ambassador to the Court of Maximilian. His account is that the seed when put in the earth grew a plant resembling a lamb, and attaining to the height of 2½ feet. It had a head, eyes, ears and all other parts of a body as a newly born lamb. It had an exceedingly soft wool, which was used in the manufacture of head coverings. It was rooted by a stem in the middle of the body and devoured the surrounding herbage and grass, and lived as long as that lasted; when there was no more within its reach, it died.

CHAPTER I

PHILOSOPHY OF FARMING

Every endeavor of mankind operates more successfully when it is based upon a sound philosophy. Farming is no different. In this chapter, I want to provide a basic philosophy of farming which should enhance the farmer's ability to achieve his basic goal of more bountiful crops.

We live in an age in which only 3 percent of the country is identified as being farmers. Yet that 3 percent must feed and clothe the remaining 97 percent. American who live on the farm are also feeding and clothing a significant portion of the world.

This is a commendable record, but we American farmers can do better. Every year we hear more stories of people who tried to make it farming and went broke. Consequently, fewer and fewer of us try to make our living off the land.

There is a good reason for the reduced number of farmers. That reason has to do with the basic philosophy of farming. To help you put this philosophy in perspective, let me begin with a story of three young men who wanted to become farmers.

Young man number one started to prepare his land. The tractor ran good. His implements functioned as they should, and his plans were right. There were some loose bolts on the tractor. He forgot his box of wrenches, so the bolts were not tightened since it was too far to go back to get them. The tractor broke down and before he could get it running again it started raining.

The delay came about because the machinery was not kept in running shape. Several times that year the same thing happened. His crop was always just a little behind and the young man worked extremely hard because his machinery was so often down. He worked hard and made a fair crop, but not a fine, big crop.

Young man number two was always a leader. He made the best grades in school and was always considered a hard worker. He didn't waste time asking experienced farmers advice, for he knew – after all, he made good grades in school. His equipment was set and calibrated just right, for he knew. Unfortunately, his seed was planted too deep and the starter fertilizer was just a little too rich for that type of land. He missed making a stand. Old experienced farmers in the area knew better, but he didn't ask them. When his crop came up late, he fertilized it to encourage it to catch up. His plants grew large and kept right on growing and he couldn't stop them. The plants were large with very little fruit on them. When the crop he had was ready to open, it rained and just kept on raining. His crop was expensive and not very good. His only excuse was that the weather just did not cooperate. The weather was the cause of his not making enough to cover expenses.

Young man number three went to an experienced farmer for advice. His equipment was ready on time when the weather permitted. He stayed in the field to see that everything ran right every possible hour that it could run. The planter was ready to go on time. The equipment was ready and calibrated when the time came. He sought and learned the practical way to make a crop and did it on time. The plants came up clean and on time. His cultivator was ready to go and he kept the crop clean on time. The insects were controlled on time and cheaply. The crop was beautiful and the effort relatively little. The expenses were small and the profit large. Why was this? Each job was done on time and looked after as experience had taught.

There was not much difference in the three young men. Each was on a par with the others. The same weather. The third young man had his crop well planned and carried out all the jobs on time. This made his work easy and profitable. Nothing can stop the third young man. In a few years he will have sufficient experience on his own to continue to do a good job so long as he continues to learn and do the what is necessary on time.

To make good crops each year, the job must be looked after and done on time. Good farming cannot

From 1820 to 1850 immigrants to this country lived in homes similar to this one.

wait around on the weather, rather the good farmer has everything ready to do the job at hand when the weather allows it to be done.

So the basic philosophy of farming presented here is really very simple. Learn from experience, seek out and follow good advice, and then be ready to act on that advice at the appropriate time. Do the job correctly when it needs to be done and you save yourself extra work and money.

Probably the hardest part of this philosophy is gaining the experience necessary. That aspect is exactly what this book is all about. I want to share with you more than sixty years of experience in raising many types of crops, but primarily cotton.

The most important lesson of my experience as a cotton farmer has taught me is that we must observe, understand, and follow nature's laws if we want to grow a good crop and become successful farmers. We must learn that nature is our friend and not our enemy. Nature's laws are always constant, they never vary. The plant responds to sunshine, food, and water in the same way every time. When we understand this we can be ready to care for the plant in the appropriate manner no matter what turn the weather may take. We must work with these wonderful laws of nature that never vary or fail. When a failure does occur we will thus realize that it is our fault, not nature's.

The plans that nature makes are wise and they always succeed. The farmer who has learned by what nature has taught him knows by experience, and this knowledge will stand as truth under all the tests that man can put it through. This knowledge gained through experience is what I am attempting to put down for farmers to use from now on. There will be so many farmers who won't make enough crop to pay their expenses. This will be caused by the farmers' not understanding the laws of nature which govern plant reproduction. But if they read and follow the advice in this book, they will understand nature's laws, be ready to obey them, and make a bountiful crop.

The goal of this book, then, is to provide access to experience and instruct the reader in the simple understanding of nature's laws. To that extent, I am trying to teach a farmer. A true teacher instructs the world in ideas and all civilization travels of the wings of those ideas. The teacher gets rich by knowing that he or she has instructed mankind to rise above present circumstances and improve his own condition. A true teacher realizes that he brought nothing into this world and he will take nothing out of it. He is satisfied that he has contributed to its development by opening the minds of those who tried to understand the basic truths that were taught. My idea is to enhance the education of the farmers of this country.

A typical cotton-picking scene from 1870-1920. It was on farms like this that the majority of cotton was raised.

As you can see from the story of the three young men who wanted to become farmers, farming is often a matter of making decisions. The farmer must make the decision as to when to plant. Too early and the seed cannot germinate. Too late and he may not have enough time to make a crop. The farmer must make the decision as to when to terminate the production of the cotton plant. If he makes the decision to continue feeding the plant and the weather stays dry long enough to harvest the added crop, he made a good decision and will reap the reward. If he continued feeding and it begins raining, and continues to rain for a long spell, then he made the wrong decision and he will see his potential profits rot in the field. Fortunately, experience can help farmers make the right decisions. In this connection farmers need to be ready to do some thinking. Farmers will have to do things just a little differently each year, for no two years will provide the same weather patterns. So the farmer will have to use his own ingenuity to accomplish the same purpose each year. A man who cannot think or cannot use his head to make things work has no business trying to farm.

A farmer to be successful must wake up. He or she must open his or her eyes and look at conditions as they are and be willing to learn. Agriculture is demanding the best brains available right now, and will continue to do so in the future. Do not aspire to be a farmer unless you are willing to stay awake and use all your faculties. If you don't use all the intellect and talents you have available as a farmer you will go broke.

Farming is not something that is done one year and then drop it for a few years planning to pick it up again later. We farm every year and take the years as they come, and profit from each experience. In sum, farming has many requirements. A farmer must be a dreamer, a mechanic, a philosopher, a student, a diplomat, and an excellent judge of other men.

The day of haphazard farming is over. If you want to produce you must know what causes the plant to produce. The plant always behaves in the same way. Conditions and what the farmer does may change, but the plant never will. This means the farmer must understand how nature acts in the plant he is growing. This truth was reinforced by Dr. Charles Lincoln, retired head entomologist at the University of Arkansas, who wrote, "Cotton cannot be forced." One must read the plant and react to its needs, just as you are now doing with this book.

This means the farmer must get the best advice available. Obtaining the best advice has moved the farmer into another era: the era of the consultant. Modern farmers must seek the advice of chemists, entomologists, and a wide variety of other specialized

This land had just been cleared of trees to plant cotton. Because of worldwide demand, all the land was planted with cotton; so from 1875 to 1920 this scene could be found anywhere along the rivers and bayous.

experts who can assist the farmer on the fine points of growing plants. Most importantly, these experts help the farmer read the plant by telling him what is going on inside the plant.

Making a good crop every year is doing a series of little things on time and in certain ways for the plant's benefit. When the plant starts to grow and develop toward maturity, it will tell you what it needs. This will have to be done on an almost daily basis. Now the way the farmer learns what the plant is telling him is to take samples and provide those samples to the consultants who know how to read the signs the plant is providing. A farmer hasn't the time to attend to all these details and take care of his business. He simply must find someone he can trust – a good consultant.

Another necessary element for successful farming is almost daily attention to the job. Farmers today have not had to work in their fields for long hours as their fathers did. Rather, modern farmers will get through with their job in a hurry, using fast and powerful tractors. They listen to their problems over the radio. Someone tells them how their plants look and how they are acting under the weather conditions. They never take the time to go into the fields and walk across them and make their own observations and decisions and form their own opinions. They get in their pickup trucks and drive through their thousand acres in twenty minutes. This approach to farming helps to explain why so many go broke and why the American farmer is a vanishing breed. The farmer needs to take time in the fields, closely examining the plants and collecting the petiole samples to send to his consultant.

The closer we adhere to the fundamentals that are known to be correct, and will be presented in this book, the greater our production will be. The way the plant reproduces has not changed, so the problem with low yields must be due to the farmers' methods. The farmer wants to make the plant conform to the methods he wants to use. The plant rebels, and the farmer cannot understand why it won't reproduce by his method. All the farmer can really do is to help the plant help itself. There is nothing mysterious about increasing cotton production, just follow the directions of common sense practices. It is really a matter of uniting scientific principles with the experience of years of farming. The method I propose can be outlined in fourteen steps:

1. Prepare the land with a three-tenths slope.
2. Be sure the ground is warm and plant on a high bed so the sun can keep it warm.
3. Use phosphorus and very little nitrogen for a starter fertilizer.
4. Use Temik on the seed or at the time of planting to kill nematodes in the soil.

A small cotton farm and its pickers.

5. Use an insecticide 20 days after the plant is up, and then at 10 day intervals two more times.
6. Have petiole analysis done to keep the food in balance within the plant.
7. Keep the early worms off the plant by using the appropriate chemicals.
8. Use insecticide in small quantities to help the squares come out at each node.
9. Feed the plant according to the results provided from the petiole analysis.
10. Put water on properly so the ground won't get soaked.
11. Use the water meter to determine the plant's moisture needs, and apply water accordingly.
12. Observe the plant daily, watching for changes within the plant.
13. When the diet needs changing within the plant do it immediately through foliar feeding.
14. Always adjust the diet so that squares will be set and bolls retained on the plant.

The above represents an outline of what the remainder of this book focuses on and explains. I have been known to make the comment, "the method never fails," whereupon someone responded, "if all other conditions are favorable, the method never fails." For example, you provide proper moisture and then a rain comes and the plant is water logged and starves for nitrogen. Although a plant is programmed to produce when fed a proper diet, days of cloudy weather interfering with the plant's manufacturing process could cause young bolls to be dropped, and so we are back to the weather excuse. But remember, I have already pointed out that the weather is really a lame excuse for not being ready to do the job the plant requires at the proper time.

Being a good farmer is a matter of doing a series of little things on time. If you have neglected the little things your crop will not be good.

1. Prepare the land on time.
2. Plant on time.
3. Cultivate the grass out of the crop on time.
4. Keep the early insects off the plant on time.
5. Feed the necessary food on time.
6. Water on time.
7. Pick it in the early fall when the weather permits.

There is one thing that stands in the way of all farmers who wish to up their production. That is doing what needs to be done on time. Everyone of us just put off until tomorrow what needs to be done today. We don't get our equipment ready in time for it to operate on the day that the job needs to be done. Getting this equipment ready always seems to take longer than we think it should. When a farmer takes short

A typical cotton-picking scene around 1900 on a family farm.

cuts and pays no attention to the "little things" that are necessary to do on time, then he is hurting himself. A farmer must feed the plant right, water it right, and control the insects right.

The only way to learn to farm is to do it. You cannot become a good farmer by only reading books. You have to take the book knowledge into the field and practice the advice given. There is nothing that will take the place of practical experience in farming. The longer you farm and the older you get, the more convinced you will be that this is true. In my opinion there is nothing better than experience when it comes to raising cotton. Remember what happens just once has to be analyzed closely to see if it can be duplicated. If I am not mistaken, the cotton plant acts the same way in response to certain conditions that it always has, and it always will. It is the farmer who doesn't understand the conditions and consequently cannot foresee the outcome.

Nature never changes its habits. Its laws never vary. Experience is the key that will unlock the door to nature's secrets.

It is necessary that a farmer have experience, and if he doesn't have it, that he knows who to ask or where to get it. The lessons experience has learned can be obtained from others, but it takes the most intelligent people to acquire it in this way. May those who are wise understand what is written here and may they take it to heart and prosper. Nature's ways are right, and when followed farmers prosper by adherence to them. Those who don't follow nature's laws stumble and fall because they ignore the fundamental truths which never vary.

Philosophy of Farming

Early cotton house and bale press around 1830 to 1900.

CHAPTER II

UNDERSTANDING THE COTTON PLANT

One thing every farmer must do, if he hopes to be successful in his endeavors, is to understand as much as he possibly can about the nature of his crops. The cotton farmer is no exception. If he hopes to make a profit from his primary cash crop, he must know all he can about the plant.

There is still a great deal to learn, in particular the inner workings of its chemistry and how it turns food into energy and translates that energy into reproducing itself, but we have begun to make some progress in this understanding. After more than sixty years of observing and studying the plant, I do know that it behaves in exactly the same way to given conditions. There is never any variation. And the key to making a profit in cotton farming is to understand nature's laws as they work in the plant and work with nature instead of against it.

Always remember, getting the bolls on the plant and getting them to stick is not guesswork, nor accidental. The plant follows the reproduction pattern exactly as it has done from the beginning of time. There is a reason for the plant putting on squares and there is a reason for the plant to keep them on and bring them to maturity. It is a matter of the reproduction habit of the plant and it works in exactly the same way under the same conditions as it has worked from the beginning, and it will always work that way.

The plant wants to make a boll at each and every node, but it needs the farmers help to do this. If the farmer will follow nature's laws, the plant will perform as he desires. The cotton plant doesn't change. It makes squares and blooms and bolls in exactly the same way as it did in the beginning. If the plant doesn't make a crop it is the farmer's fault, not the plant's.

The key to knowing what the plant needs and helping it perform to its best potential is for the farmer to get out in the field and observe his crop. The farmer must get out of his pickup and walk in the field, turn the stalk back and see how it looks. How many squares are there? How many bolls? How long are the internodes? After he has made his observations, he needs expert help in determining exactly what action to take. Sometimes a plant's food requirements can be observed through the naked eye, but not often. A great deal of damage has occurred in the plant before we can see it. So he needs to depend on a good agricultural consultant and on a chemist trained to identify what food is within the plant and what food it needs to perform properly.

Over the years of being a cotton farmer, I have learned that there are six main rules to be observed to make a good crop:

A common bale press from about 1850 to 1900. The bale press screw was hewed from a tree. This screw when pulled by a mule caused the bale to be tied.

1. The plant must be looked after every day.
2. It must be fed a diet conducive to its reproduction habits.
3. It must be kept clean from competing vegetation.
4. It must be given moisture when it gets thirsty.
5. But it cannot stand water around its roots, so the land must be sloped to provide good drainage.
6. It must be protected from insects.

All these elements must work together for the farmer to make a good crop. If any one of them is not observed, it hurts the whole program. Weed control, insect control, water, and food all must work together. The main problem we face in this program right now is that we know less about the plant's food needs than any of the other areas, but we are learning more all the time.

The cotton plant is one of nature's great plants. It has been this way from the beginning. Man has not changed it. He has, by selection obtained a better plant – one that will mature a few days earlier – but the process has been slow and only slightly beneficial. Desirable characteristics must continually be sought and selected. Even so, the basic plant will never change. Thus, the farmers job is to:

1. Take petiole samples
2. Feed at the proper time.

3. Feed through the leaves because this permits a more timely job to be done.
4. Pay attention to the manner and timeliness in how the feeding program is carried out.
5. Be humble and try to learn from nature, for she is the one who holds the answer to the problem of production.

Good cotton farming begins before the seed is put into the ground. In fact, preparation of the land is one of the six key factors to successful farming. It's been found that cotton produces best in soil which has a 3 tenths slope to it. Cotton in land with less than a 3 tenths slope is not as healthy and does not get all the food and nutrients from the soil that it should. The land should also be broken deep to give the plant's roots the ability to go deep into the soil and provide footing to support the stalk and all the bolls it will produce. One of the things our big powerful tractors have done is to pack down the soil, forming a hard pan which is difficult, if not impossible, for the roots to penetrate. So we need to plow deep to break up this hard pan when the soil is dry.

The preparation of the seed bed is also important. I have found that it is best to plant the seed in a bed which has been hipped up pretty high, providing clear furrows and allowing the warmth of the sun to get into the soil and to the seed. The high bed also makes it easier to plow to control the weeds and to make a channel for irrigation purposes.

Today, most cotton farmers are using triple treated seeds for planting. Triple treated seeds are seeds which have been treated to make them resistant to disease and insects. There are a number of equally good varieties of cotton seed available. Generally, the selection of a seed is a matter of personal preference, or what may be popular in a given area. In the past some effort was made to provide particular types of plants. When people had to separate the fiber from the seed by hand, farmers were interested in finding a plant which would allow the seed to be pulled out easily. Now gins do this tiresome work, and the ease of separating seeds from fiber is no longer a deciding factor in selecting a seed. Plant breeders over the years have come up with a wide variety of cotton plants. At one time breeders developed a cotton plant that would grow tall so a person would not have to stoop so much when gathering the cotton. Now with mechanized pickers the tallness is no longer desired.

I generally try to plant the seeds in the early part of April, as soon as the danger of the last frost is past and the weather permits the machinery to get into the field. I have learned from experience that early planted cotton has shorter internodes than late planted cotton, and the desirability of shorter internodes, which has been mentioned before, will be further explained as we discuss the development of the plant. Cotton seed will germinate in from 4 to 10 days, sometimes a little longer, depending on the weather conditions. In cool weather it takes the seed longer to germinate. In hot weather it will germinate in four days. It does not seem to injure the plant for the seed to be planted in mud. Five weeks after planting there will be early squares on the young cotton.

When planting, I have found it is best to try to have the stand as close to one plant per foot of row as possible. This is conducive to a stronger stalk which will hold more bolls. It also contributes to faster blooming and better boll setting. In addition, it seems to cut down on the extent of boll rot. I have observed that the cotton stand with between one and three stalks per foot of row will yield more pounds of seed cotton per acre. In such a stand bolls are more uniformly spaced throughout the field and on the stalk. These plants also hold up the boll load much better, with less laying down in the row, we also find the crop is not ready to harvest as early as where it is better spaced out. Where the plant has space to grow, it also allows the sun to get down to the lower bolls and help them mature faster. In thick crops the plant tends to shed more of the lower bolls.

How the mule press operated: As the mule pulled the screw, (hewed out of wood) the cotton was compressed with a hole in the middle. This system was widely used until the advent of the steam presses.

The first two leaves will appear on the plant in about two weeks after it breaks the crust. After that, the plant will add another node each six days. The first fruiting branch appears at the seventh node and the early squares will appear on these branches. In a regular six day pattern, the plant will add another node and more squares every six days. There will be one leaf to feed each square. When that square has become a mature boll the feeder leaf will drop off by itself.

Always remember, the time to make the crop is in the first part of the life of the plant, or in the first 75 days. It emerges in 30 days, puts on squares in 50 days, and should be blooming in 75 days. If the farmer has adjusted the food in the plant as it should be, he will have a crop blooming on top by July 15th. It is best for the young plant not to have too much food available. If they have too much food they will grow too fast, get too tall, and consequently will not produce as much fruit because they put all their energy into growing instead of producing. When the plant starts to set fruit and gets its habit set, then feed it and it will produce lots of bolls. The plant can be kept short and forced to put out short internodes by anyone who is willing to follow instructions. The key is knowing what food to give it and when, and this is why the chemist is so important to the modern farmer. The stalk cannot be forced to get shorter after it is already too

The cotton was ginned by hand gins in the upper story of this building. It was then carried in baskets to the press on the outside where it was pressed into a bale. Circa 1830 to 1900.

big, you have to start on time and keep it small and not let it get too large.

When the internodes are 3 to 4 inches long, the plant is growing too fast. If the internodes can be kept to between 1 and 2 inches during the plant's early growth and set a square at every node, then the stalk will not get too large. When the petioles are long, as well as the internodes, it denotes rapid growth and that the food is out of balance. This probably means the plant has too much nitrogen during its early growth. If the growth starts to get too fast, it can be controlled by feeding the plant sugar.

Bear this in mind, anytime the cotton plant gets clean and the fruiting branches are coming out holding all the fruit, it is then that the plant is ready to produce. If it doesn't hold all the fruit it is making, the farmer is at fault, not the plant. We have found that all the squares can be made to stick and will make large bolls if the plant is fed and protected properly. This means keeping the competing vegetation out of the field as much as possible, giving the plant the proper food at the proper time, keeping the insects off of it, and providing moisture when needed.

The internodes must be kept short in order to make a maximum crop. If a square is at every node and the nodes are 6 or 7 inches long, the farmer is still in trouble as far as making a big crop. The internodes denote

Since there were no roads or railroads, it was necessary to build the mills where the cotton was grown. In North and South Carolina, Georgia and Alabama the cotton mill machinery was hauled to the cotton fields by mules and cattle. Imagine moving heavy machinery today as they did in 1896. The job got done!

how fast the cotton is growing. There should be a square at each node and by mid-June the internode should not be over 2 or 3 inches. If it is longer than that the growth is much too fast for good production. If there is not a square at each node, there is a reason, and the farmer had better find out why. It could be improper balance in the food within the plant or it could be insect damage. Notice the cotton that has internodes of 5 or 6 inches. That cotton also has very large leaves. The leaves are much larger and the petiole stem is much longer than the shorter cotton with smaller leaves. These large plants are a dark green as compared to the lighter green of the smaller plants, and the darker plants reveal too much nitrogen in their early growth and will not make as much cotton.

The program discussed in this book is intended to keep the plant shorter but stockier and to encourage it to set squares at every node. This enables the plant to better support the number of bolls it will make. Food regulation is the main source of controlling the growth of the plant. Another regulation of growth comes from the herbicide used to help control weeds. When we spray for weeds we also retard the growth of the cotton plant somewhat, but not enough to injure it.

In growing cotton, it is so important for the plant to develop and hold its early fruit. This can be done by providing the right food during the growing season.

All plants wish to produce and they will do so if they receive the food that is necessary for them to do their best. There are three main reasons for the plant to throw off the early squares: Insect damage, food deficiency, or lack of moisture. If the farmer attends to these three things during the growth of the plant, he can help the plant hold 95 to 100 percent of the squares if puts on and bring them to maturity.

The plant grows on an exact schedule, speeded up or slowed down just a little in relation to the hot or cold weather. A given number of days after it breaks the crust and starts its life it will put on its first joint above the first two leaves. The branches that produce fruit will start about the seventh node up from the first two leaves. It takes about six days for each node to grow. It will start making squares about 36 days after it comes up. Warm weather will speed this up a little.

Cotton is a hot weather plant. It loves heat. The hotter the weather, the faster the plant reproduces. Heat also speeds up maturity. But during hot weather the plant uses the available food faster. In July, 100 degree weather is just what the plant needs to do its best, along with ample food and water. Get the food adjusted in the plant in June and the plant will produce fruit in July. If the food is not brought into balance in June then the plant will go into excess growth in July instead of fruiting, and then it will have to fruit

This was a typical good road crew of convicts that built and maintained the first roads all over the South. Before the bulldozers, motor graders and even graders pulled by a string of mules, the roads were built by prisoners with shovels. Notice the guards with shotguns, if a prisoner tried to escape, he would be shot. Circa 1915 to 1920.

in August and September. The problem with that is the winter weather may not allow the crop to be picked. It is much better to make the crop early so it can be harvested as early in the fall as possible. The truth of this method is that the action of the plant from day to day is always in exact relation to the food, moisture, and heat the plant has available to it at a particular time.

I believe that farmers will produce more cotton by controlling the insects early. The best crops come by saving the early bolls. It appears to be best to keep the plant growing slowly and to keep the three early insects off of it. The three early insects, which can cause the plant to shed its early squares, are thrips, hopper, and plant bugs. I control them with what I call "square set" which is a commercial insecticide which kills these three insects. I use a mild dose on the plants twenty days after they come up and then go back with another dose ten days later. Then use additional doses to fight the insects as the plant has need for it. Later in the summer some insecticide used with the food foliar fed program should be used.

By keeping the early insects off the plant, the cotton will set a square at each node and will hold those squares, because the insects are not destroying them. This means the plant will have lots of squares by the first of June. The June blooms will be mature bolls which can be picked by the middle of September.

By the middle of June the squares are coming out at each node. There are numerous insects which want to live and reproduce just as the cotton plant does. The insects lay their eggs and the eggs hatch and start eating on the very tender growth of the plant. The cotton plant is their food and their place of reproduction. In turn, there are bugs that eat these insects, and they are present because there is lots of food available for them. The great question every year is what to do about this condition. It would be nice to destroy the enemy bugs and save the friendly bugs,but the insecticide treats them all equally. If we did a 100 percent job of destroying the enemy bugs only, the friendly bugs would starve to death anyway. It is my opinion that more cotton will be produced by controlling these enemy bugs before they can do a lot of damage. In this way the early bolls are saved and a larger crop is produced. If the insects are not controlled regularly and continuously the crop will be significantly smaller. Where the insects are not controlled, farmers find a large number of missing bolls and soft bolls. The missing bolls have been shed. The soft bolls are the result of weevil and red spider damage. Soft bolls are so named because they never get firm, and thus never mature or open up.

Most farmers have accepted the fact that too much nitrogen in the early part of the plant's life can force the plant to grow too fast and too big before it starts to bloom. In most of the cotton growing region of the United States, because of the limited growing season, that rapid growth is not desirable. Cotton will live the year around if it is not killed by cold, and in more moderate climates, such as Mexico, it is used as an ornamental tree.

The early bolls are usually the best and most profitable bolls of all. So do what is necessary to help the plant bloom early and hold all these bolls. There is a way to speed up the fruiting of the fine varieties of cotton our plant breeders have supplied us. That is to limit the amount of nitrogen at the time of planting to just enough to provide healthy growth. Then as the plant develops the bolls on the bottom and begins to square at the top, put on lots of nitrogen to feed these rapidly developing bolls. Fly the food on. Apply water by irrigating, if needed. Then if the plant seems to have more nitrogen that it needs and is starting more vegetative growth, control this by applying small quantities of sugar, 2 or 3 pounds per acre.

Keeping the food in balance helps the plant put on squares. Keeping the squares on the plant is largely a matter of controlling the insects and worms, not allowing them to knock the squares off. The effect of insects on holding squares has been verified by Dr. Jackson Mauney, the USDA plant physiologist at the

Western Cotton Research Laboratory in Phoenix, who writes, " ... most of the damage in the first six to eight weeks of squaring is caused by insects." Another reason the plant will abort squares is due to the improper balance of food within the plant. When the nitrogen level is too high, the plant will abort squares and use the nitrogen for growth rather than reproduction. It requires no great effort on the part of the plant to produce three blooms instead of one, but it takes food in the proper balance for it to bring three bolls to maturity instead of one. The main reasons for not having a square at each node are:

1. The food is out of adjustment.
2. Insects damage them.

The plow may knock some of them off with the machinery as we plow the crop to help rid the field of grass and weeds. Not very many are destroyed this way because once the square develops it sticks on pretty well unless the insects get it or the plant aborts it for lack of a proper food balance.

The importance of having the food within the plant "in focus" or in proper balance becomes apparent during the peak reproduction period in August. At that time the crops that were brought into focus in their early stages (June) will have lots of bolls on them. The plants will not be too large, and they will have a square at every node. There is no mystery as to what makes

Carrying seed cotton to the gin to bale. The cotton in this wagon was probably 1/2 cotton and 1/2 seed. It took about 1,100 lbs. of cotton to make one bale.

the square set and the bolls reach maturity, it is a matter of keeping the food in proper balance within the plant. The plant must be brought into focus in the 8 to 10 inch stage so that the internodes will not grow too long. I have observed that if the food is right within the plant the squares will develop at least two weeks earlier than other cotton.

I have recently come to the realization that I judge whether cotton is ready to produce a big crop by its color. Is there a way to determine when two colors are the same? The eye can tell, but we don't know why, and our judgment can play tricks on us. If the food within the plant is correct its color will be a healthy light green. Not dark green, nor blue-black green, but a lighter shade of green. The color of the plant depends on the food it is getting. Dark green means too much nitrogen. Light green is a plant with food in balance. Yellow is a plant that has been water logged. Rain can water log the soil (especially if it doesn't have the proper slope) then the plant can't get food out of the soil because the water has replaced the food. A dark blue-green color in cotton indicates that it maybe has been treated with Pix to retard the growth. I prefer to use sugar to control growth because the sugar helps the plant alter the food balance naturally, whereas Pix seems to cause the plant to slow its growth by injuring it. Anytime the plant is injured, even a small amount, its production will be reduced.

Dwight Lincoln of Arkansas conducted an experiment a few years ago in which he compared cotton fields which were treated exactly the same except that one had Pix and the other did not. Dwight reported that the Pix field produced a half a bale of cotton per acre less than the control field. The blue-green color is an indication of the injury to the plant. Anytime the plant is injured, even a small amount, its production will be reduced.

In July the plant works as hard as it can to produce itself. It is utilizing all its energy to make squares and bring them to bloom. Then it is utilizing all the food it can find to bring the bolls to maturity. During the height of boll development notice the large number of five lock bolls. A five lock boll is the result of the proper food balance within the plant. The lack of nitrogen at this time will cause the cotton plant to quit making squares and shed its fruit. There may be plenty of moisture in the ground, and the insects may be under control, but if there is not enough nitrogen to feed the squares and bolls, the plant will shed them. The demands on the farmer are very great at this time to get the food to the plant. I have found that applying urea dissolved in water every five days, according to the chemist's instructions, has been an effective program for getting the nitrogen the plant so desperately needs to it. This solution is applied right onto the leaves where the plant can absorb it more quickly, and put it to work more effectively.

Food is the key to why the plant will set and hold its bolls. It is my belief that we can begin feeding the plant through the leaves (foliar feeding) when the time is right. By the first of July the plant can be adjusted inside with the proper food and it will begin setting fruit and continue on as long as we wish or so long as the time permits. We must provide the proper food in the correct proportions. Many factors have a bearing on production. Many of these can be controlled to a degree. After the plant gets up, is cleared of grass and weeds, gets to growing with the food in the correct balance, and sets the squares at every node; then the plant is ready to be fed to help it hold all the fruit and bring it to maturity. This will usually occur around July 1, some years a little earlier, other years a little later, but the first of July has generally proved to be the correct time.

The food in the plant can be adjusted much more rapidly and efficiently with foliar sprays than by root feeding. By mid-July one can see how fast the squares are coming on the plant, and the plant is getting large, so there are plenty of places for it to put squares. This large plant and all the squares require a lot of food and

Hauling bales of cotton from the gin in the country to the cotton mill close by.

water. The chances are very small that there is enough food in the ground to feed the large stalk and all of its fruit. It needs additional food, and it needs it now. The practical way to get the food to the plant is to feed it through the leaves. The crop response to nutrient sprays is more rapid but more temporary than soil feeding. Thus, the farmer must set up a regular feeding program.

We don't want to wait until the cotton grows all it wants to before it starts fruiting. We want to speed this process up through nutrition to make it mature more quickly so that it can be harvested before the fall rains. We don't need a longer growing season to make more cotton. We need to encourage the plant to make more bolls each of the days that it is in the peak reproduction period.

The cotton plant that has been kept clean (free from weeds) from the time it came up is ready to reproduce itself by the first of July. For the next 70 days this plant will try its best to make as much cotton as it can. The farmer must give it the food and water it needs to do this. The plant really needs close watching for the next 6 weeks. If the food has been adjusted properly, the plant mostly needs sunshine and water. Any rain received can cause the plant to change, but the farmer can adjust the food to meet that condition.

It takes about 60 days from bloom to an open boll of cotton. In North Louisiana we cannot expect blooms in late September to have enough time to mature, but on rare occasions they do. The 60-day time frame breaks down as follows:

3 weeks from square to bloom.

4 weeks from bloom to mature boll.

3-4 weeks from mature boll to open boll.

The first killing frost usually comes around November 7, so squares formed by August 20 still have enough time left to become a mature boll before the first frost. The farmer must get sufficient food and water to the plant during this time to help it bring the bolls to maturity. After that the boll will open on its own.

During the month of July, the farmer must be particularly alert. Careful observation of the crop is essential. He needs to maintain effective insect control. He needs to get water to the plant at the right time. He needs to provide plenty of food to keep the plant in balance and to feed all the bolls it is making, but he also needs to beware of the possibility of the plant going vegetative. To go vegetative means the plant starts growing again and aborts the young squares. If the plant is loaded down with fruit and the food balance is maintained, there is no way the plant will go vegetative. And loading it down with fruit in July also means it will bring the bolls to maturity as much as two weeks ahead of other cotton crops.

As a rule, by July 5, or soon after, the plant's pattern should be set. Then the stalk can be properly fed the food it needs to hold the fruit during August and bring the bolls to maturity. Then picking can start by September 15 and be in full swing by October 1. This is the time of year when it is good to observe all the cotton fields in the area. The cotton fields that are beginning to lap the middles are usually dark green with large leaves and stems from 6 to 9 inches long and the stalk is 45 to 50 inches high. As a rule there are not many squares or blooms on this cotton yet, but there is an abundance of dark green leaves throughout the plant. This means the food in the plant is out of balance. There is an abundance of nitrogen and insufficient phosphorus and since there is not much fruit, this nitrogen is going into the stalk growth. This type of cotton will continue to grow tall and thick until nature adjusts the food balance itself in the latter part of August or the first of September. Then the leaves will become smaller and the petioles will get shorter and the plant will take on a lighter green color. The square buds will come out all over the top of the plant and will be in clusters. If the insects are kept off and enough food and water furnished, and if the frost delays long enough, it will make a bumper crop. But if conditions are not met, there will be very little to harvest.

What we attempt to accomplish with the program we are describing here is to get the stalk to making before July 4. It should be in full bloom by July 15 if we have any hope of making a good crop. In contrast to the cotton described in the preceding paragraph, the cotton grown using this system will have blooms all over it by mid-July. Where the food within the plant is correct, the blooms will be all over the stalk, on top and on all the sides. The color is a light shade of green. The internodes are between 1 and 3 inches long and the plant itself is about 30 inches high. The chemist, through his tissue testing of petiole samples, tells us the plant is on the low side in nitrogen and high in phosphorus. This means the plant is in a position to hold all the squares. These squares will develop into bolls if the farmer provides enough nitrogen (urea) through the leaves, and gives the plant sufficient moisture.

For all practical purposes, all the squares that have bloomed on or before July 15 will be open and ready to pick by September 15. All the squares that bloom after July 15 can be picked when they are open, probably in late October or early November. Many more squares can be held on the stalk after July 15 than before, but it requires ample feeding to bring these bolls to maturity. Remember to feed the nitrogen to the plant in July to set the bolls and not cause too much growth. Keep the ground damp during July and August, the plant will really produce if this is done.

As the time approaches (July and August) when the crop's demands for water will have to be met by irrigation, remember to always use a lay-by after opening up the off rows over which the water will run. The term lay-by means one last application of herbicidal oil to hold down the grass and weeds.

By early August most cotton will begin to "cut-out". The meaning of this term is the plant will stop both growth and reproduction and will begin to open up the bolls it has. This occurs when the available food in the plant is about gone, or it has stressed for moisture. The reason a plant will stop making and "cut-out" is that it is without food or moisture. Sometimes the insect damage can cause this. Nature does not cause the plant to "cut-out". Food and moisture are the primary reasons the plant will stop. When this occurs, the leaves begin to turn yellow and nearly all the young squares and bolls fall off. The leaves then shed and the bolls that are left will pop open, beginning on the bottom and then up the stalk.

The plant cuts out and quits for one or more of three reasons:

These wagons are lined up at the warehouse, waiting to store their bales of cotton. Each wagon held from four to six bales.

1. The plant gives out of food to feed the young squares and bolls and it stops.
2. The plant runs out of water to supplement the food and it quits.
3. The leaves are injured by insects to the extent that they are useless to the plant and thus it quits functioning.

By this stage of the plant's development, the main insects to cause concern are aphids, white flies and spider mites. If the cutting out occurs in September it could be a help to the farmer, since it causes the bolls to open up so they can be picked before the bad weather sets in. However, most of the time cutting out is not a beneficial development.

In late August of 1982, the cotton over most of North Louisiana was beginning to cut out. What happened was the food began to give out and the plant was still quitting even though it had enough moisture. The plant's reaction to its necessities is not limited to any locality or to a particular farmer. The plant acts exactly the same way to food and moisture conditions. It will throw off squares and bolls where there is not enough food and moisture to bring them to maturity. Of course, the plant performs best when there is ample moisture, the insects are kept off, and the food is maintained in the proper balance. The two main reasons for losing bolls in the top of the plant are dry weather and lack of food.

By mid-August the plant should be literally blooming all over. Additional blooms are coming out at the side of the stalk and at the base of grown bolls. The internodes are getting shorter and the squares come out in clusters. From this time until September 15, or about 30 days, the cotton plant can produce bolls, after this time it will be too late for the bolls produced to mature. The days in August begin to get shorter and this makes the nights cooler. The cotton plants speed up their reproduction process and set squares at every possible place on the stalk. All the squares produced before August 20 will make mature bolls of cotton if the insects are kept off, there is ample water and enough food is furnished.

In the middle of August the squares come out in clusters. The internodes are very short, and the squares come out on the side of the stalk, and at the base there are many mature bolls. By this time the crop has been up and growing for more than 100 days. It has been blooming for 50 to 60 days. It will continue to grow and produce just as long as the warm weather holds, there is sufficient food and moisture, and the insects are controlled. The stalk is big and capable of holding lots of bolls, but the bolls need food to grow to maturity. August showers can be most beneficial, for the extra moisture will help more bolls to stick. The main reason a plant throws off fruit is a shortage of food and water.

The cotton plant can make more cotton faster during September than at any time of the year, if the weather will allow it to be picked. When the plant is fed through the leaves (foliar feeding) it will continue to grow and develop all the fruit it has on it. The natural process of the plant's development at this time causes the internodes to become shorter until the bolls will almost be touching each other. The real key at this time is to furnish the food necessary to hold all these bolls. The cotton plant will continue to grow and put on squares and mature the fruit just as long as there is food and water and the insects are controlled. The plant can be kept producing until frost if the farmer desires.

By mid-September the bottom leaves shed naturally. The bolls continue to fill out normally. This is exactly as it should be if the fall weather holds out and rains don't rot the bolls. September 15 is the last day a white bloom can make a good boll of cotton before the frost. The reason is that is takes 3 weeks from bloom to full grown boll, and then another week for the boll to harden to the point it is ready to pop open. This will bring the date to October 15. The normal frost date is November 7 which gives the boll three weeks to get mature enough to open before the frost which would

A steam engine would haul the logs to the mill and then haul the bales to the docks where the steamboat would take them to the world market.

sour it if it is not yet open. All blooms on the plant by September 15 will make a boll of cotton, which is in keeping with the earlier statement that all squares on the plant by August 15 will make a boll of cotton.

Up to this point, I have alluded to the need for moisture, but have not discussed it very much. Never let cotton stress for water. Once it has suffered moisture stress it never gets back into its normal stride of production. The surest way to avoid this problem is to use moisture meters at strategic locations in the field. The moisture meter can tell you what the level in the soil is and you can get water to the plant before it starts to stress. The cotton that has moisture never stops producing. I have noticed in the past that the last of August and the first two weeks of September the squares get closer together and seem to stick better if there is enough food and water. Another indication of the need for water can be seen in a field I was shown in 1982 where it was not watered on time. That field did not make more than one and a half bales of cotton per acre because of the water problem, whereas the cotton next to it made 3 bales per acre.

Too many farmers suffer because they depend entirely on nature to provide adequate rain fall. We cannot depend on rain, even in the South. So we have to develop a system of irrigation. The object of irrigation is never to let the plant wilt or suffer from lack of moisture. If it does suffer it will shed its fruit and start all over again. A good farmer is prepared to supply what the plant needs when it needs it, and this often means irrigation.

Most farmers think they have it made if they get the rain, but this is not necessarily so. They may get the rain, and the cotton will still quit because it doesn't have enough food. Rain changes the food balance within the plant and thus stimulates growth and the new growth spurt can cause it to abort squares. It take 5 or 6 days for the plant to undergo this change. We have a chance to control this with foliar feeding because this system gets the food the plant needs into it within 12 hours. Wet weather reduces the nitrogen content of the soil. The cotton plant uses up the remaining nitrogen rapidly. It will set early bolls and then when later squares and bolls come on there won't be enough food to hold them on the plant, and they will be shed for lack of food. Another problem with rain can occur when the plant has been fed nitrogen and that feeding is followed by a rain. This rain can cause other foods to come forward as well as extra nitrogen. This upsets the food balance in the plant and encourages more growth which means longer internodes and the aborting of some squares.

The final consideration regarding rain has to do with the normal fall rain that interferes with harvesting

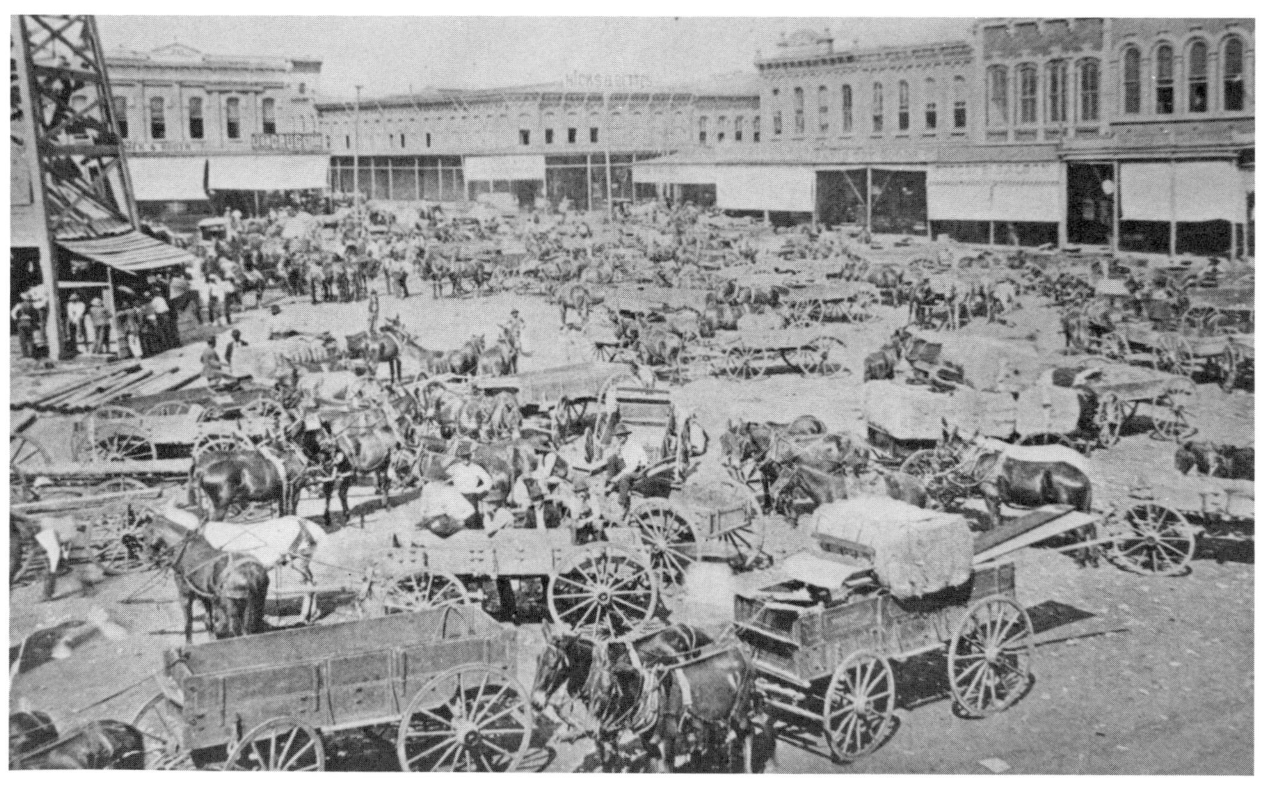

A typical Saturday evening in the fall in the Cotton Belt. The streets were jammed with people buying supplies and visiting.

the crop. In 1981 the rain started on October 6 and continued throughout the remainder of the month. The rain damaged the bolls which had been made in the month of August when they were trying to open in October. August was the month the plant had fruited the heaviest. I estimate the loss of the potential crop that year at 50 percent. When the boll starts to open, if we get a lot of rain, the fiber never fluffs out and we get what are called hard locks. In addition, all the moisture increases the likelihood of boll rot, and all the bolls on the ground, due to the weight of the rain on the plant will rot even faster because the ground holds the moisture longer. Everything will rot in wet humid weather, including mature bolls.

After the plant blooms, it naturally sheds the bloom as it starts to make the boll. It takes four weeks after the bloom to have a full-grown boll. After the boll reaches full size, all it does is open up, no more food is required by that boll. All the nitrogen fed to the plant will then go to the young squares. Each leaf that comes out at the time the young square develops is responsible for the welfare and development of that boll. When the boll reaches maturity, the feeder leaf will fall off naturally. But for some reason, if the boll is not there the leaf will remain indefinitely. On a well-nourished plant, the feeder leaf will drop off when its work is done, but if there is too much nitrogen the leaf never sheds.

Each boll of cotton requires lots of nitrogen until the boll gets four weeks old. At that time it is mature and it has to dry out and open up. It will take about four weeks after it is mature before it opens and is ready to pick. The time it takes to open varies according to how hot and dry it is. The fiber in the boll gets dry and expands, and this causes the boll to literally pop open. This process is enhanced, in a healthy plant, by the fact that once the boll reaches maturity the feeder leaf drops off, because with the feeder leaf gone more sunlight can get to the boll and it will dry out faster.

A four-lock boll is one which has four divisions in the boll – you can see four supporting prongs under the boll. A five-lock boll has five such divisions and prongs, the bottom of the boll resembles a star. The assumption is that five-lock bolls have more fibre in them, but we don't know that for sure.

As harvest time approaches, I pay particular attention to the advice provided by the chemist. For example, in 1979 I applied 15 pounds of urea per acre and one-fourth pound of methol on September 10. The urea fed the late blooms and the methol killed the weevils. Then on September 29 I flew on 8 pounds of sugar per acre and another quarter pound of methol. The sugar was used to slow the plant's growth and bring it to

Another Saturday evening scene between 1875 and 1920.

maturity. On October 5, I flew on another 6 pounds of sugar and quarter pound of methol. On October 25, I defoliated the crop. All this was done at the recommendation of the consultant chemist I was using.

The time to defoliate is when 50 to 75 percent of the plant is open. Generally, it can be picked a week after it has been defoliated. There are several brands of defoliant available. We do need to pay some attention to proper timing for defoliation, because to defoliate too early may cause small bolls since they were not allowed to reach full maturity.

Most farmers in this area try to get two pickings from the crop. I generally try to pick around September 15 and again about a month to six weeks later. The first picking does not appear to do any damage to the remaining bolls. When the cotton on the bottom of the plant is open and ready to pick, we should go ahead and get it. It rarely pays off to wait for the whole crop to open up, usually bad weather or some other problem occurs.

Cotton seeds for a long time were considered a waste product, something to get rid of. Many early gins were located near streams or rivers. They needed water for the oxen, mules and horses the farmers used to haul the cotton. Many of these ginners would dump the seed into the rivers and Bayous to get rid of them. Eventually laws had to be passed to prevent this sort of pollution. I have long had an interest in disposing of trash from gins, and in preventing fires and dust irritation to the lungs of the workers. I am happy that I was able to solve the problem of gin trash. The waste products of cotton gins are now being reclaimed instead of burned. A secondary gin will gin the motes, as the material is called, which are then dried and cleaned again with the clean cotton being baled to be sold. Sometimes the spinners can use this material, and some types of paper can be made from it. The extracts of sticks, green growing matter, and seeds is mulched into a compost pile to sell to gardeners. We have also found other uses for the seeds. With the utilization of the seed for food for man, think how many acres of land this releases for other purposes. When man started using magarine made from cotton seed, and oil made from cotton seed for cooking instead of butter, the number of cows needed to make butter was reduced.

Understanding the Cotton Plant

In small towns, cotton was purchased by buyers and shipped by rail to the warehouse.

CHAPTER III

SOIL PREPARATION AND PLANTING

Here are a number of factors which go into producing a successful cotton crop, not the least of these is having the soil properly prepared. There is an old saying about a woman's work is never done, the same adage applies to successful farming. Almost as soon as the farmer gets the last crop harvested he has to start planning for the next crop.

One of the key factors in producing a bountiful cotton crop is having the land sloped properly for good drainage. Let me emphasize again and again, drainage comes first, then break the land. If the land needs to be sloped, do that in the fall, then break the land before Christmas. Use a plow that cuts deep enough to break up the hard pan, the packed soil a few inches below the surface that has been packed hard because of all the traffic that has gone over it – tractors, disks, trucks. Then hip it high in March.

If the land is prepared properly, excess rain water will run off rather than stand in the field creating a situation which would lead to rot. The proper slope for the land is a 3 tenths slope. This means the land falls 3 inches for every 100 feet. The soil needs a slope to it so that rain water will run off and not water log the soil creating a situation where the roots cannot get the fertilizer that is in the soil. If it is not prepared to drain well, it does not make any difference when and

Buyers hauled their cotton bales to the world market by means of river steamboats.

how much fertilizer is used, the crop will not be good. Poisoning the insects does not do that much good if the land is not drained. Having the best cotton picker money can buy won't do any good because there won't be much to pick. The best crops are made on a high bed on well drained land.

Drainage is essential and comes first in all our efforts. The best drained land yields the best crops. Drainage makes it possible for the crop to grow early in the Spring so the plants can withstand the cool, wet weather. Drainage allows the soil to produce its best because it won't get water logged. I have checked the comparative results of properly prepared land versus land that was not properly prepared. Wherever there was a good slope the cotton was much better. If any part of the cotton was planted on land that did not have a 3 tenths fall the size and development of the plant was noticeably different. The stalk will be much smaller and will not have the vigor that the cotton on the slope will have. The food goes out of the ground faster. Where there was a good row fall and the plants were left on a high bed, the cotton developed and produced much better. Where the land slope was not over 2-1/2 tenths slope the nitrogen got out of the ground and went off somewhere because it was not in the plant according to the petiole analysis. The stalks in such places were small. Finally, the cotton in land with less

There was always plenty of fun for all. A good telephone – short or long-distance calls.

than 3 tenths slope needs more frequent watering than the 3 tenths slope land.

In 1981 a comparison of the production figures of cotton in properly prepared land and in land with less than 3 tenths slope all the cotton received exactly the same treatment in all respects, the only difference was the slope of the land. The results are as follows:

Cut 1	3 tenths slope	6 acres	12 bales produced.
Cut 2	1½ tenths slope	6 acres	7 bales produced.
Cut 3	1½ tenths slope	6 acres	5 bales produced.
Cut 4	1½ tenths slope	6 acres	5 bales produced.
Cut 5	2 tenths slope	6 acres	8 bales produced.
Cut 5	3 tenths slope	6 acres	12 bales produced.

The 12 acres with a 3 tenths fall produced 24 bales of cotton, or 2 bales per acre. The 24 acres with less than a 3 tenths fall produced 25 bales, or just over 1 bale of cotton per acre. This difference amounted to $300 per acre or potential lost profit of $7,200.

Obviously, then, for cotton to do its best it has to have a good fall to the row. It should also be planted on a high bed to help it get off to a good, early start. This type of soil preparation enables the cotton plant's feeder roots to warm up early, and it helps the soil hold the food it contains by preventing denitrification, which occurs when the soil gets soaked.

Once the land has the proper slope, the hard pan needs to be broken up. To do this, I use the Big-Ox plow and sometimes go down as deep as 18 to 25 inches. The reason for breaking up the hard pan is that this enables the cotton plant's roots to get deep into the soil and form a solid footing for the large stalk that will eventually develop. The deep plowing also allows air to get into the soil, and the plant's roots need some air as well as water in the soil.

My experience in 1982 helps to explain the need for deep plowing. I formed the land in December of 1981 to take out a hollow section in the middle of the field. After the land moving work it rained a lot and we were not able to get into the field to work it more. All the heavy equipment used to form the land packed it down even more, and due to the rain I wasn't able to deep plow to crack the hard pan. Consequently, the 1982 crop was forced to put out shallower root systems and the packed soil did not allow the roots to breathe as they should have. The crop never seemed to get off to a good growing pattern. The factors to remember in proper soil preparation are:

1. Be sure the land has a 3 tenths slope.
2. Break it right plow deep to crack the hard pan, and to bury the weeds. Then it will be easier to cultivate in the spring and early summer.
3. Too much use of the disk tends to pack the land below the 8 inch level. This makes it difficult for the roots to sink down. This packing of the soil also

harms the water drainage capability of the soil, since packed soil won't absorb moisture as well as loose soil. The packing of the soil 8 inches below the surface is what I have been referring to as the hard pan.

Since one of the advantages of proper soil preparation involves helping control weeds and grasses, it has been found that the use of a herbicide, after the land has been given the proper slope and deep plow, is also desirable. In the fall of 1980 we flew on Treflon and then disked the field with the versatile tractor. We then hipped it up at once and let it stand over the winter. The following Spring the land was hipped up again, drug off, and planted. There were no morning glories or similar vines in that field. On land adjacent to this field, where herbicide was not used until the spring, there were plenty of morning glories and weeds.

After the land has been through the winter, we go back to the field in March. At this time I will chisel plow several times until every sprig of grass is separated. Then I go over the field with a large disk-harrow, which breaks up plowed or rough land by means of a series of disks arranged at an angle with the line of draft, to smooth the land out. I will follow that with the disk-hipper, which is a cultivator attachment with two series of disks arranged to throw the top soil around the roots of the crop or into a hill, to form a high bed. If

A scene in the farming country from 1880 to 1920. "Cutting the Pigeon Wing" was a well-known dance – ask any old-timer and he will describe it. This is a part of history that we must not forget.

necessary, I will hip it up several times. Right behind the hipper I run the large Logan roller to firm up the bed and make it just right for planting.

The reason for having a high bed is to get as much heat to the plant as possible. The ground has to be warm for growth to take place, and when the bed is high the sun can get to it more efficiently and warm it up so the plant will grow much better than it will in a low, flat bed. The disk-hipper can be set to either throw dirt onto the hill or to throw the dirt and weeds away from the row.

While I do believe that a high bed is necessary for a good cotton crop, I would put in a word of caution about excessive use of the disk. Too many trips over the field tends to pack it down, forming a hard pan which is not conducive to a good crop. In addition, I have noticed that disking the land always makes it dry out early, and thus makes it harder for the young plant to grow. It is better to use the disk as little as possible.

The soil is the place where the seed and then the young plant will get the nutrients it needs to develop into a healthy cotton stalk. This means that often the farmer will need to provide some preplant nutrients, and then use a starter fertilizer. In our area, one of the needed nutrients is sulphur. Some people have a regular schedule for adding sulphur to the soil, such as every five years. I tend to prefer to test the soil to determine what it needs, and then add the needed nutrients according to what the test results indicate. I test the soil in the fall at 6 inch and 18 inch depths. I send the soil samples to Al Lengyel in Phoenix, Arizona who can tell me what, if any, nutrients to add. I will have another soil test done right after I finish planting.

After the cotton is to a stand, the farmer will be able to tell where the land was properly prepared. Where the cotton is the smallest the land is the flattest. The water comes over this cotton and leaves debris all over the small plants. In addition to having the proper slope to the land and hipping it into a high bed, it is helpful to put out a little starter fertilizer when planting to give the young cotton the best possible start.

A starter fertilizer is a small amount of fertilizer with a high phosphorous content. Young plants need phosphorus to get off to a good start. However, when the ground is cool, as it usually is in the early spring, the plant can't get as much phosphorus because cool land doesn't release the phosphorus it contains as rapidly as warm land. I recommend the use of liquid phosphorus as a starter fertilizer because it is easier to handle. Rich land causes the plant to speed up square setting, but the same high land will get the plant growing too fast and making long internodes and

After buying supplies and a trip to the blacksmith shop on Saturday evening, you headed back home with a fresh barrel of water from the spring. Circa 1900.

petioles and very large leaves. If the land is allowed to stay with the nitrogen content higher than the phosphorus, then the young plant will begin to abort the early squares and this in turn encourages it to grow again. So I encourage the use of phosphorus as a starter fertilizer and little or no nitrogen.

In summary of this section on soil preparation, the farmer should provide a high bed using a disk-hipper in the fall. He should put out a starter fertilizer in the bed at the same time. Then leave the row alone until spring. If necessary, hip it up again before planting, and then drag it off lightly with the Logan Harrow. It is a good idea to incorporate Treflan in the bed with a rotary hoe. When planting time comes, drag the bed off lightly again, but leave it high. Plant on the high bed. Then, do not plow or cultivate at all until after the vegetation is killed on each side of the plant for ten inches (a twenty inch band). Do not cultivate until after the drill is thoroughly clean – that is free of all competing vegetation and weeds.

I believe the method of soil preparation I used in 1978 is the correct method, and I would like to describe that method at this time. We tried to prepare the land in beds before the winter rains started. The row was hipped up high and then the Logan harrow was run down them to make them uniform and Treflan was put out with the rotary hoe. The bed was checked for correct height and the Logan roller run over it to assure a smooth surface. The bed needs to be fixed right even if it takes extra trips over the field for this to be done. The Treflan will clean the bed if the top is smooth because the herbicidal oil will cover the grass on a smooth surface. The herbicidal oil is so much better and cheaper than chemicals, because the oil does not hurt the crop as much and thus delay the good start.

By bedding the land up before Christmas and not disturbing it there will be lots of grasses coming up through this old bed, but the seed bed will be a better place for the cotton roots to develop. After the bed is cleaned with the oil, usually two trips will do the job, then a plow can be run down the middles to clean them out.

We must get rid of the Johnson grass. That is the weed that really bothers us the most now. Treflan will make the Johnson grass lay down so it won't grow in the drill, but it won't kill it. The herbicidal oil does suppress the Johnson grass until we can get back and spot spray it with Round-up.

If the land is prepared right, the farmer won't have to plant nearly as many seed to obtain a good stand. Planting a seed every 3 or 4 inches is more than enough to assure him of a stand. One thing we do have to be sure of is that the temperature in the soil is warm enough to allow the seed to germinate. To do this, the

AN AVERAGE FARMER — Many people believe that it was slave labor that produced most of the cotton production in this country. This is far from reality. It was the average American farmer who raised from 100 to 300 bales of cotton, doing most of the work themselves with their families, that built the cotton economy.

farmer is well-advised to purchase a thermometer. There are soil thermometers available, but I have had equally good success using an ordinary cooking thermometer. I take temperature readings at 2″, 4″ and 8″ depths. Seeds will not germinate until the ground reaches 65° at 2 inches in depth. Then to be sure of the available food in the soil, I have a soil test run again at planting time. Once the cotton is to a stand, I will begin taking petiole samples from the plants to determine the food level. The important place for the food is in the plant, more so than what is available in the soil.

When the time comes to plant, have the planter ready, don't start fixing the planter the day you need to plant. Plant the crop correctly, for you are going to have to work it. If you don't plant right you will have to work harder to get the ground right and keep the weeds out.

Plant your cotton as early as possible (to give the plant as much growing season as possible) on a high bed. The reason for a high bed is that the sun warms the ground from the top down, and a high bed warms up earlier than a low bed or flat ground. I like to plant in late March if possible, but usually plant in early April. Some farmers wait until May. A cotton farmer will plant whenever he thinks he can get away with it, but in any event he should wait until the last frost is past.

Sometimes planting may be delayed because of an unusually cold spring. If the spring stays cold for a long time, as it did in Louisiana in 1978, it is better to wait until May 1 to plant. There are times when cotton planted as late as the middle of May will be the best cotton of all. The problem with planting as late as May is that getting a good crop will take the same number of days to mature as the April planting date, the cotton plant always performs in exactly the same way, so the farmer will have to pray for a good fall and late winter to harvest his crop. Generally, however, it is a good idea to plant as early as the weather will permit and take advantage of as many good growing days as possible.

Knowing how many seeds to plant involves a lot of guesswork because we don't know how many of the seeds will germinate. When the cotton gets to a stand we like to have one stalk per foot of row. The only way to get the number of plants down to one or two per foot of row is to plant them that way, but such a planting method does not guarantee that kind of stand. I generally plant the seeds between 2 and 3 inches apart. I prefer the 3 inch distance if all the seeds germinate. One year, 1978, I planted three inches apart but not all of it came up and I ended up having to replant in May.

A picture of some of the early settlers who made this nation. They raised what they ate and used the bales of cotton (protected under the tree) as their bank account.

I usually plant between 9 and 12 pounds of seed per acre, which means putting a seed down between 2 and 3 inches apart. I use an acid delinted seed planted at a depth of 1-1/2 inches. The farmer wants to plant the seed deep enough to germinate well. Usually between 1 and 2 inches. In heavy soil we plant a little deeper than in sandy soil. The farmer will have to use his own best judgment, there is no set way for planting, either in terms of space between seeds or in terms of depth.

I use a six-row drill to plant my cotton. I also prefer to leave every-other off middle open, this provides some space between every-other row which allows the sun to get to the plants more efficiently. At the time planting takes place also add square set to the drill. I use Temik as a square set, which helps protect the young plants from the early insects, especially thrips. It is a good idea to use Temik when planting the seed. Where Temik is used, the plant sets earlier. I do not know why, but it works every time.

There are some changes in cotton planting coming in the future. I believe the future of cotton lies in broadcast planting of cotton. The reason we don't do more of it now is that we still don't have a good broadcast harvesting machine.

Once the cotton is to a stand, it needs to be cleaned. Plow and clean the whole field when the crop is small.

Get it perfectly clean and keep it that way. The farmer can plow as often as he desires and not injure the plant, the important thing is to keep it clean from competing weeds and grasses so all the available nutrients go to the cotton plant. If the soil has been properly prepared, as I have discussed earlier in this chapter, it will probably not be necessary to plow so frequently. But the farmer will have to plow enough to keep the competing vegetation under control. If the weeds are not present, the cotton plant will grow just as well without cultivation.

I think the idea of cultivating very shallow is correct. By not distrubing the roots any more than necessary, the plant gets along much better.

The roots of the plants grow better and produce more when the seed bed is firm. The older farmers fifty years ago and older before scientific farming took over, have a saying—Cotton won't start making until the roots hit hard ground. Experience over the years has shown that this is true. In the days when crops were cultivated with one mule and one plow, the end of the rows where the mule turned around, the land was never broken or the soil loosened up by the cultivator around the plant. The best stalks of cotton with the most bolls on them were found on these turn-rows.

One problem the farmer must be prepared to face is the problem of needing to cultivate in the face of a

A country doctor and his helper in front of his office in Arkansas about 1900. These doctors grew with the country and looked after the health of its people.

long siege of rain. I faced this situation in 1981. After a lot of rain, we plowed where we could. In some spots the ground was crusted over on top and wet underneath. After we plowed the ground dried and became very hard. Then the second time we tried to plow, the blades would not go into the ground. The remedy is to plow the ground the second time as soon as it dries and before it gets hard. This will force the farmer to wait longer in getting over the field the first time he plows, but that won't make much difference because where the ground has never been plowed it won't get so hard.

Cultivation is necessary only to kill grass and weeds. This does not help the crop to produce, and if the field is clean of grass and weeds it is a waste of time and money to cultivate. Sometimes, cultivation is cheaper than using herbicides, but probably both will need to be used to control the weeds. Cultivation should be done as shallow as possible. The blade should be set just deep enough to destroy the vegetation but not disrupt the cotton roots.

A deacon and his saddle mule. Taken in 1910.

CHAPTER IV

CONTROLLING WEEDS

Having prepared the soil properly, and planted the seed on a high bed, the farmer should be in a good position to watch his cotton come to a stand. However, he can't just sit around and enjoy the spectacle of nature making a crop. He will have to be constantly on his guard to do what is necessary to help the young plant reach maturity. One thing he will have to do to accomplish his goal of a bountiful harvest is to keep the weeds and grasses under control so that all, or as much as possible, of the food available is going to nourish the cotton plant and not competing vegetation.

The farmer must decide early to do what is necessary to help the cotton get off to a good start. That means controlling the weeds to keep the rows clean. In 1982 I had trouble with the grass and weeds in my crop because they were not controlled before they came up and were not controlled when they were small. Consequently, the weeds got too large and I had to expend extra time, effort, and money to clean the crop. In addition to this waste, we didn't produce as much cotton because of the early weed problem.

Frequently, if the real truth were known, this is what happens. The crop is planted in a half-way fashion – some seed deep, some shallow, some in the moisture, and some in the clods. The seed is not treated and the preplanting herbicidal oil is not applied. After the crop begins to come up, the middles look bad, so a plow is run down the rows. The plow rolls some clods under the plants in the drill. When the herbicidal oil is applied under the plants to control the grass and weeds, the clods protect many of the small weeds and keep them from being killed. As a result there is grass and weeds growing in the drill. This situation is really more troublesome and cuts down production. And it is the farmer's fault for not doing the proper job of cleaning it out when the time was right. So he will blame the weather or bad luck when the trouble really is himself in not using his head and doing each step in a timely way.

By planting our cotton properly and then not bothering the smooth surface until we have the drill perfectly clean, we can relieve all the grass and weed problems connected with the cotton crop. The procedure in a nutshell would be this:

– Drain the field perfectly, build up a high bed early; plant on this high bed and then run a roller over it to smooth the top perfectly.

– When the cotton plants get large enough to go under, clean this smooth surface with herbicidal oil.

– Two weeks later clean it again with oil.

– Then run the side winder down the middles and clean them and the crop will be in fine shape to produce with no grass or weed problems at all. This method will work every year, everywhere.

Wherever the plant is not free of other vegetation in the drill, the plant has a chance to be injured. The drill should be cleaned of weeds and grass in the month of May. The vegetation needs to be kept out of the crop so that all the food and light will go to the cotton instead of the weeds. Where the weeds are not controlled early and properly the farmer may find his yield cut by as much as one bale per acre.

Herbicidal oil will wipe the grass and weeds out if the farmer desires to do so, and the oil will not injure the young cotton plant. Sometimes, in the process of getting the spray rig properly adjusted I have gotten the oil on the under side of the cotton plant leaves. Those leaves were shed, but where the third and fourth leaves were coming out the plant continued to grow and did all right. The proper time to apply the herbicidal oil for the first time is when the cotton plants get to be about 4 inches high.

Generally, the herbicidal oil is sprayed on the drill and on the stems of the cotton plant. Cotton has a protective oil on its stem that does not allow the chemical to affect the cotton, but it kills morning glories and causes Johnson grass to lay down. Morning glories and Johnson grass are the two biggest headaches we have in Louisiana insofar as controlling weeds is concerned.

DALE CARNEGIE

Author of "How to Win Friends and Influence People" } Says--

One day a farm hand in Oklahoma drove to town, went to the post office, made out a money order, and sent for some books. As a result of that your life will be a bit changed.

The hand's name was John D. Rust.

The books he sent for dealt with automobile mechanics. He earned the money for them by picking cotton. At night, when the family had gone to bed, John D. Rust sat up reading his books on mechanics. One night the farmer came in, with an old-fashioned nightgown sloshing around his ankles, and bawled him out. Said that he was wasting coal oil, and told him that he would go crazy if he sat up at night reading such books.

The farm hand said he would go crazy if he thought he had to pick cotton all the rest of his life.

That farm hand had a vision, absurd as it seemed at the time. He believed that a machine could be invented which would pick cotton. The more he thought about the idea, the more deeply he believed in it. He sent for other books. One was a set called "The Library of Original Sources."

His education had been neglected, for he was an orphan and had to make his own way in the world. But these books were a start. The first rung on his ladder. He was able to lift himself out of the farm hand class and get a job with a threshing machine company in Hutchinson, Kan. Meantime, he was working on his cotton picker idea.

He worked out a spindle with teeth in it. The spindle would pick the cotton all right, but he couldn't get the cotton off the spindle. One night in April, 1927, he went to bed worrying about the problem. It seemed impossible to solve. Suddenly he remembered that as a boy when he had gone out to pick cotton early in the morning, the cotton had stuck to his fingers on account of the dew. Then he said to himself, "Maybe the lint will stick to steel if the steel is damp." He fished around until he found a nail, wet it with his lips, and twirled it around in some absorbent cotton and lo! the cotton wrapped itself around the nail. The secret of the cotton picker was solved then and there.

John D. Rust went into partnership with his brother Mack, and together they worked out the world-famous Rust Cotton Picker. The first cotton picker came off the assembly line Nov. 4th, 1938. I have a letter from John D. Rust, which says that he had ten pickers in operation in 1938. This year he expects to go into production on a modest commercial basis. What that farm hand worked out will revolutionize the cotton growing industry, whether for good or bad only time will tell.

The important thing is that John D. Rust used his spare time for achievement. He set a goal, never deviated from it, and has done something that will influence almost every person in the world.

START SCRAP BOOK

Many readers are cutting out these columns and pasting them in scrap books so that they may review and study them as a text book on success. Why not start your Success Scrap Book now?

Baby Kenneth Johnson is touted in London as the future world's heavy man for he is less than two years old and weighs 80 pounds.

An overdose of vermifuge, santonin, makes a person see all objects that cross his vision as either yellow or green.

Dale Carnegie newspaper article

When the plant is large enough we try to spray MSMA in a way that allows us to go up underneath the cotton plant. This will get the MSMA more directly on the grass in the row, and does less damage to the cotton plant since this puts the MSMA on the stalk rather than on the leaves. The cotton stalk contains a protective oil which the MSMA can't penetrate.

The mixture of herbicidal oil to water that I use calls for 1 gallon of MSMA per 12 acres mixed into 10 gallons of water per acre and sprayed under 60 pounds of pressure. I will put this on and then come back with another application of the same mixture in 5 to 7 days. The MSMA will cause Johnson grass to turn brown and lay down in the row. This mixture does not harm vigorous cotton plants, but sometimes it will kill sick or weak plants.

To kill morning glories, I prefer to use Cortoran. This herbicide is used at a ratio of 1/2 pound per acre mixed into 10 gallons of water per acre. As with the MSMA, I apply it once, then come back with a second application in 5 to 7 days. I have found that I can apply both Cortoran for morning glories, and MSMA for Johnson grass in the same mixture. Cortoran is especially effective against broad leaf weeds (such as morning glories), while MSMA works on grass.

The intervals between applications of the herbicidal spray is a matter of choice. I usually put the second

First practical cotton picker in Louisiana on Dan Logan's farm in 1936

application on five days after the first. As I said above, I have found that I can mix MSMA and Cortoran together and apply them at the same time. I have also found that mixing a little soap in with the herbicidal oil helps the mixture spread more evenly. For this purpose I use VPG spreader at the ratio of 1 gallon of VPG to 200 gallons of water.

The farmer does want to be careful about the application of the herbicidal oil to kill the weeds and grasses being done too frequently. When MSMA is used too frequently and too late it is not as effective against the weeds and grasses, and there is a stronger likelihood of its having a deleterious effect on the cotton plant itself. This means the cotton is weaker and can't produce as well as it should, and the weeds are healthier and thus taking more of the available food from the already weakened cotton plant.

Nonetheless, the farmer may need to make more than just two applications of herbicidal oil to keep the weeds under control. The important thing is to keep the cotton as clean as possible during the month of May, and if this means a third or fourth trip across the field to spray the oil, then he should do so.

The last application of herbicidal oil I call "layby". This is my term for the last application of chemical put on the ground to control late season morning glories in particular. This is put on just before the feeding program starts, and is usually done in conjunction with the final cultivation of the crop in preparation for opening up the ditches for irrigation. This usually occurs in mid-June. Remember, it is wrong NOT to put out "layby" to kill the grass and vines, because they take such a heavy toll of the crop. This is particularly true in the case of foliar feeding, since the vines will get so much of the food.

In addition to using herbicidal oil to control weeds and grasses, the farmer will need to plow (cultivate) the crop several times to further reduce the effect of weeds. The herbicidal oil can help us control the weeds in the drill, the plowing helps control the weeds in the rows.

It is important to plow and clean the whole field when the crop is small. The farmer should get the crop perfectly clean and keep it that way by whatever means he has available. You can plow as often as you wish and not injure the plant, just keep it clean as best you can so the cotton plant can use all the available food rather than having the weeds steal the food from the crop.

I learned with my crop in 1978 that it may be best to plow really shallow. Doing so still keeps the weeds down and does not disturb the plant roots. In any event, I try not to plow or cultivate any more after July 4. It is about that time that I do my "layby" plowing, which means that this will be the last time I make a pass through the field for weed control purposes. From this point on I assume the cotton is strong enough to take over any competition from the weeds and I put my effort into providing all the food it needs. This is also the time to prepare the field and rows for watering purposes.

One aspect of cultivation that I learned by the hard teacher of experience is that we should not plow when the ground is still wet. Now I have known that for a long time, but one particular year, after a long siege of rains, we were so anxious to do some plowing and clean up the fields, that we went in after a crust formed but the underneath was still damp. After the ground dried it was extremely hard, and later when we tried to plow again, the plow would not penetrate it. This means we have to wait longer to plow when the ground is wet, but ultimately the ground will not be so hard if we do wait.

I am not so sure that we really need to plow all that often. This is particularly true where the soil has been properly prepared. But we are also helping ourselves by the fact that we have more and better herbicidal oils and chemicals available to deal with the weed problem. So we can spray for the weeds more effectively and have to plow less freuently. I have become convinced that frequent trips with the tractor over the field is doing more packing of the hard pan than any benefit those trips are providing to the plant. We do need to keep the weeds down to prevent competition for the available food, but we can do that with chemicals and occasional cultivation to keep weeds out of the rows. In coming years, cultivation will be used less and less.

Controlling weeds and grasses is the farmer's primary responsibility to his cotton crop during the month of May. To control this excess vegetation and keep his crop as clean as possible he will need to be prepared to spray a herbicidal oil on the weeds and plow frequently to clean them out of the rows. In addition, the plowing can help the cotton plant's roots breathe more easily to help it get the best possible start. This way the plant can devote its later energy to production of bolls.

With a clean crop, we can now devote our attention to providing the necessary food and water for the plant. The layby plowing is our last trip through the field and is done with the water needs of the crop in mind. Soon we will be taking steps to be sure the crop has plenty of water to enable it to absorb the food we will feed it. These ideas are presented in the following chapters.

Controlling Weeds

Dan Logan's two granddaughters, Susan Logan and Beth Logan, help survey the land so that it can be cut to a "3 to 1" slope for the best production. They made the survey, drew it on graft paper, and set the stakes so the land could be cut to grade.

CHAPTER V

GETTING FOOD TO THE PLANT

I like to think of the cotton plant as being similar to a family. A family has children, and each of those children need plenty of food to continue growing and ultimately become mature adults. The cotton plant puts on lots of squares, and those squares need a lot of food to have a chance to become mature bolls of cotton. The farmer's job is to see that the plant has plenty of food to enable the plant to bring all those potential bolls to maturity.

Feeding the cotton plant really begins at planting time, when we put down some starter fertilizer with the seed. A starter fertilizer is a very small amount of fertilizer with a high phosphorous content to give the sprout some food to help it get a good, healthy beginning. Young plants need phosphorus. When the ground is cool, as it usually is at the time we begin planting cotton, it does not release much phosphorus, so the young plant needs a little help. As the ground warms up it releases greater quantities of phosphorus. Liquid phosphorus is a good starter fertilizer because the liquid form is easier to handle. A small amount of preplant fertilizer does not hurt, but a lot does. We need to rely on a good chemist to tell us how much preplant fertilizer to use. He can tell us this information based on his soil analysis. Later on, when we get into the regular feeding schedule for the plant, the amount used is not so important, but it is necessary to provide the food at the right stage of growth for good production.

The important thing to observe about feeding the plant in its early growth stages is to be sure the food is in balance. If the food within the plant is in balance, the plant is not growing too fast. The food within the cotton plant needs to be adjusted so you won't have an excess of one food and too little of another food. If the nitrogen within your plant is lower than the phosphate, the chances are great that the internodes are not too long and that there are lots of early squares. In the early stages of the cotton plant's development, the farmer should work to keep the balance between nitrogen and phosphate within the plant in the proper ratio. Once again, a good chemist is most useful in providing this information. In addition to the major nutrients being in balance, we must also keep an eye on the balance among the other minor foods: calium, boron, zinc, calcium, sulphur, iron and manganese. If the food within the young plant is in proper balance, the internodes will be from 2" to 3" long and there will be a square at each node. When the food is not in proportion, usually it is too high in nitrogen, the internodes will be well over 3" long, often as much as 6" and the plant will abort many of the squares it started to produce. It seems as though every time we get the food within the plant in the proper proportion the plant will cause the square buds to come out at each node. Then all the farmer has to do is feed those squares and keep the insects from eating them and in a few short weeks we have an open boll of cotton to pick.

The plant knows whether or not it is going to grow big when it only has 8 leaves. It knows whether or not the food is out of balance before the fruiting branches come out on the plant. We must find out from the chemist what the plant is going to do, then we must make the necessary food adjustments so it will fruit instead of growing big. The way the chemist can tell us is through the petiole analysis. I use the fourth leaf from the top of the plant and take petioles from a number of different plants. Petioles, when analyzed, give us as much information about the health of the plant as a blood test tells us about the human body's condition. Taking a weekly petiole analysis is right and will prove out over the years. Someone with knowledge and experience must advise us. That is the role the chemist can play. I begin taking petiole samples after the plant has six leaves, and continue taking samples on a weekly basis.

It is my belief that the first three samples are the most important. The fourth and fifth samples will show you whether or not you have made the necessary food adjustments and the plant is on the road to proper production. The square buds should be coming out at each node and the internodes should not be over 2 or 3 inches long at the most, the shorter the internodes the better. Sometimes when this exact method of production has just been started, the fourth, fifth and sixth samples will still show the need of further adjustment. As your production goes up you will pay more attention to doing just what you are told to do in the first three samples and chemists recommendations. Food adjustments are made according to what the petiole tests reveal. Even then there is some guess work involved because the food we apply may not all get into the plant. The best advice is to let the plant grow as nature intended. We do that by feeding it the food it needs rather than what we think it needs. Find out what the cotton plant needs by taking petiole samples and having then analyzed by a competent chemist. Petiole analysis can be useful if we learn to use it wisely.

An important concern to the farmer in the early stages of the cotton plant's growth is controlling the size of the stalk. A lot of farmers are using Pix to slow down or stop the plant's growth. Pix will slow down the growth, but it does so by injuring the plant. I prefer to use sugar to help control the growth. Sugar works naturally within the plant to retard growth. It does this by altering the chemicals within the plant rather than injuring the plant. Another reason I prefer the use of sugar is that it is cheaper to use than Pix. Sugar can also be used near the end of the growing season to speed maturity.

Sugar is a carbon, as such it causes the nitrogen-phosphate ratio with the plant to change. Some people use boron, which causes a like change. I prefer to use regular granulated sugar as a foliar spray. When the internodes get too long, sugar sprayed on every three or four days for three or four times helps to reduce the internode growth. The sugar will enter the leaves and bring about the nitrogen-phosphate ratio adjustment. I learned in 1980 that the effect of sugar on the growth of the plant will not last over four to five days. Thus the plant may require several applications. Normally I use four pounds of sugar per 10 gallons of water, and put out 10 gallons of water per acre. The amount and frequency of the sugar applications may vary according to the chemists recommendations. In about ten days you should notice a change in the plant, the internodes will be getting shorter and the new leaves will not be as large. If the plant has too much nitrogen in relation to the phosphorus, then the squares won't all be set and the internodes will get long. This can be controlled almost every time by the proper use of foliar feeding with sugar.

I have mentioned foliar feeding in connection with the application of sugar. Perhaps I should now explain what foliar feeding is. By foliar feeding I mean applying the food to the leaves of the plant in a liquid spray form. A good ground spray rig is the key to proper foliar feeding.

I advocate the use of foliar feeding because it is the most efficient way to get the food into the plant. One pound of nitrogen foliar fed through the leaves is worth eight pounds put in the soil. The University of Arkansas reports that it takes 150 pounds of nitrogen per acre to get 2 bales of cotton per acre, but this assumes putting the nitrogen on the ground. The same amount of nitrogen foliar fed amounts to 1200 pounds per acre. In addition, foliar feeding gets the food into the plant much faster. It takes 3 days for the plant to acquire and use the nitrogen it needs through the soil, but foliar feeding gets the food into the plant in 12 hours.

A combination of root and leaf feeding would probably be the best, and in fact I use both methods, but all the food the plant needs to produce a bountiful crop can be fed through the leaves. Once the plant is into it regular production schedule, usually in July and August, the farmer should be into a regular schedule of foliar feeding, usually every five days.

The most satisfying way to get food into the plant is through the leaves rather than the roots. A plane can be used or a ground rig that is capable of going over the top of the plants and spraying on this food. It does not make much difference how the plant gets it, just so it gets the food when it needs it. Gene Woodall and Dick Maples both agree that is is impossible for the cotton plant not to set squares if the method of feeding I am advocating is followed. If there is not enough food in the plant to feed all the bolls, the chemist can detect that and tell us how much to use. Foliar fed nitrogen can get into the plant in less than one day's time and stop the square shedding.

It is necessary to maintain a steady but not too rapid growth during the fruiting period. This can be done with foliar sprays much faster and more accurately than through ground feeding. The lower leaves that feed the lower bolls will be benefited by foliar nitrogen to develop their respective bolls, and the nitrogen foliar fed to these leaves should make larger and better bolls. I recommend using low biuret urea for foliar sprays because it dissolves into the spray effectively. The best measurement for determining whether this system works is the crop yield, and foliar fed crops have consistently been high producers.

It seems as though foliar feeding is possibly better than feeding through the roots in several ways. It is faster; it can be controlled more easily; and it is economical. Foliar and root feeding could work together very well if we study them more. We will have to go into analyzing the plant for complete food earlier and feed the necessary deficient elements through the leaves so that it will get the food immediately. By the time our eyes see it the damage has been done, we must rely on a consultant-chemist to tell us what to do.

The foliar food gets into the plant very quickly. You can begin to notice it in one day, and in two days it is easily noticed, and the third day anyone can see it. The fast action is what is needed for it can stop the plant from shedding its fruit, where food is short, almost at once. This fast action is good in another way; it acts fast and quits fast. In September, when you wish cotton to stop making and mature, you can feed the late squares and young bolls without getting the late growth that always plagues the farmer. The nitrogen won't be there to cause it to grow.

The early life of the cotton plant is a dangerous time to feed a lot of nitrogen because nitrogen will cause the plant to grow tall and to abort most of the young squares. This fact is attested to by Dr. Joe W. White of the LSU Extension Service who writes, "Too much nitrogen and not enough phosphorus and potassium creates strong vegetative growth at the expense of setting fruits". Dr. White continues, "Flowers will be formed under these conditions but will drop from the plant without having set any fruit. Also, too much nitrogen in the late summer and fall will keep plants succulent for too long which often results in winter injury." We know that nitrate put out early definitely causes the plant to delay setting fruit.

By adjusting the food balance within the plant the stalk will grow slower, the internodes will be shorter and more squares will come out and consequently more bolls will mature. This means the farmer should follow the advice of his chemist and the advice given here, get the plant's food in balance, avoid too much early nitrogen. But, Dr. White tells us, "after many crop plants have developed to a point where early evidence of maturity exists, for example, when early forming fruits are nearing a mature size, an additional, but light, application of nitrogen is usually a good practice". What the farmer wants to do is to feed the cotton plant in June to bring all the food elements into proper balance. Petiole testing will reveal what is necessary for that balance. June feeding involves nutrients other than nitrogen - phosphorus, potash, zinc. We try to go into July with the plant in balance and then feed it the nitrogen necessary to promote square development. Too much nitrogen in June will promote stalk growth but retard square development.

In June we put in enough nitrogen to keep the plant healthy, but not enough to encourage it to put on a new growth spurt. This balance of feeding the nitrogen needed for a plant to build tissue and sugar to keep the plant in a fruiting pattern must be continued during July and August, at the time the plant is flowering best. Nitrogen is the great fruit producer in a plant. It is also a growth factor. When we are too early in applying nitrogen, and do so at the wrong time, we tend to cause the plant to grow faster rather than fruit.

Another problem with nitrogen that often plagues the farmer is rain. Wet weather can cause the nitrogen to get out of the ground and this will make the plant suffer for food during July and August (the months the plant is making a crop). With the nitrogen gone, most of the squares will be shed after blooming. If the fields are too muddy to run the ground rig, then the farmer will have to fly on the nitrogen, but the important thing is to get the food to the plant when it needs it.

During July and August, when the plant needs plenty of nitrogen to feed all the squares, it is preferable to use low biuret urea. It is popularly called L.B.Urea, and if used properly does not burn the crop, and, as I have mentioned before, it mixes well in water for foliar feeding purposes. Urea is 46% nitrogen. The cheapest, safest and most practical way to apply nitrogen is to feed it through the leaves using L. B. Urea as recommended by a competent chemist. This can be fed early in the morning, as long as the pores of the leaves are open, or at night, if the machines are running. Night feeding will be most beneficial for the pores of the leaves are open and the food will go into the plant at once. This is also a good time to include the insect poison in the nitrogen at no extra cost, and it seems to do a better job when they are mixed.

The experienced farmer can tell a lot about how the plant is doing by its appearance. If the food is correct within the plant, its color will be light green rather than dark green. Dark green indicates too much nitrogen and a tendency toward stalk and leaf growth rather than reproduction. As mentioned earlier, the internodes and petioles will be under three inches long. Finally, in July and August, the healthy plant will have bolls and squares all along the stalk , and very few squares that have been shed.

If we want to make bumper crops, the important thing is to feed the plant the proper food at the appropriate time, as advised by the chemist. This will cause the square buds to come out, then feed the squares the necessary food for them to develop into full grown bolls. The late bolls can be as large as the early bolls if enough food is furnished.

Important Note:

I am writing about the herbicides and insecticides that I have used. Some of these will not be available in the future. Regulations change from year to year. Consult your county agent for reliable brand names.

CHAPTER VI

CONTROLLING INSECTS

Probably the major reason we farmers don't get the yield from our cotton plants that we hope for is insects. It is much better and cheaper to kill the insects when they are young and before they can do much damage than to wait until they can be seen. By the time we see them they have already done much of their damage. We really don't know how to keep the insects off the plant in the first place, so we have to control them from the outset.

Many of the agriculture schools and experimental stations are advising farmers to wait before using insecticides because the insecticides cannot distinguish between good and bad insects, they will both be killed. By the time the farmer sees good insects such as lady bugs in his field, the bad insects have already been doing their damage to the plant, knocking off the squares. Every square that is hurt means another boll of cotton lost. The principle rule for insect control is to control them early with a systematic application of insecticide.

The system of insect control that is good begins at planting time. Use the systemic chemical (referred to as square set) at the time of planting. The systemic chemical gets into the plant so that when the insects begin to suck the cotton plant they are poisoned by the chemical in the juices of the plant. This early insecticide helps to control thrips, plant bugs and flea hoppers which attack the early squares.

In addition to the systemic chemical used at planting time, experience has shown that after the plant has been up for 20 days a light application of insecticide (square set) should be used to kill these early pests, and then in 10 more days use it again. This approach generally controls thrips, plant bugs and flea hoppers quite well.

In order to do an efficient job of controlling the insects, it is important that we know something about the insects and their nature. Thrips are the first insects to attack the cotton. Thrips will go after the young cotton. Hoppers attack the early squares, and along with the plant bugs will attack the cotton when it recovers from the thrip damage. Hoppers and plant bugs go after the young squares put out by the plant. Tarnish plant bugs seem to migrate to cotton from the wheat fields. They attack the early squares, but are more aggressive than hoppers. The boll worm usually appears in late June or early July. It will eat everything in sight, squares first and then the bolls. Red spiders usually appear in early July and attack the plant leaves. White flies will appear in late August and September. The white flies also attack the leaves and tend to cause early shedding of the leaves so the cotton plant will not bring all its bolls to maturity.

The cost of controlling thrips is relatively small when compared to the return the farmer will get for having done so. For about fifty cents an acre the thrips can be controlled, and it can be applied either by air or with a ground rig. Thrips can be controlled with Azodrin because it is systemmic and can also be used to control red spiders and white flies. For thrip control apply Azodrin on the 20th day after the plant is up and again ten days later.

Why are the thrips, flea hoppers and tarnish plant bugs always present? No one can answer that, but they are always there when the plant starts squaring. Since the farmer wants to produce a good crop, he must control them. Twenty days after the crop is up kill them, and then in ten days do the same thing again, and the plant will be well on the way to making a good crop.

Most farmers are willing to give up the early squares to save the later ones. This approach costs them as much as one-half of their crop. Poison for hoppers and plant bugs before they get any young squares. Clean them off by the calendar, just 20 days after the crop is up. Don't worry about whether they are there or not, just clean them off; they will always be present.

As mentioned earlier, Azodrin a synthetic pyrethyroid, is preferred. It generally is used as directed on the label, except that less is put on with each application than directed, but use more frequent applications. Once the spray is started it is probably best to spray each 5 days, this usually means 12 applications.

It is believed to be better to save all the first squares and blooms rather than let the insects get them. These first bolls are larger and you get more for your poison dollar than you do later on. The plant will not start making cotton or setting fruit until after you start poisoning the insects. The farmers who wait until mid-July to poison are probably the farmers who are not prosperous.

It is probably best to begin the systemic treatment for boll worms when the boll worm eggs are on the plant. This usually occurs in late June or early July, but in 1981 the first eggs were found on May 30. The cotton must be watched every day. This means getting out in the field and looking at the plants. The boll worms will appear about the same time every year, give or take a few days. The boll worm moth with always appear in great numbers before the boll worms appear. The moth has to lay the eggs. When the moth appears then the young worms will be hatched three to five days later. The worms begin eating the tender parts of the plant. If the poison is timed right most all of them can be killed. Begin poisoning when the moths appear and then continue the program every five days. The first generation of worms won't hurt the farmer too much, but the second generation can really do severe damage. The systematic poisoning approach is designed to kill the first generation so that the second generation can be more effectively controlled, but the farmer cannot let up on the poisoning program.

One year I became quite complacent about having controlled the insects. None was seen, so I skipped one poisoning time. That was a mistake, because then the insects got a good start and I had to go on a three-day poison schedule to regain control. Once into the systematic poisoning program the farmer should stick to his schedule.

Boll worm eggs will be laid all over the tops and sides of new growth on the cotton plant. In three or four days the eggs will hatch and there will be worms all over the stalk. Controlling them is a timely job. Kill them when they are small and going from bud to young squares and then to young bolls. If we wait until they get in the boll the poison will not get to them. The full life cycle of the boll worm is 20 to 21 days. We want to control the first generation of boll worms as effectively as possible because it is the second cycle of boll worms that really clean out a crop.

Do not be fooled by the boll worms, or an apparent lack of them. They will swarm in overnight and lay eggs all over the plant. It has been discovered that mixing some sugar in with the insecticide when it is applied seems to help kill the moths before they can lay their eggs. The mixture that has been successful calls for 4 pounds of sugar, 1/10 pound of Azodrin and 10 gallons of water per acre. The best reason is that this mixture works, for moths are attracted to the sweet poison and tend to drink the spray.

Changes in the weather can have an impact the farmer should be aware of and take precaution against. Insects seem to multiply faster in cool weather. In addition, cotton is a warm weather plant; the hotter the weather, the happier the cotton. The cotton does not grow as well in cool weather, but the insects are more plentiful; so the farmer needs to keep up his systematic program of insect control. In warm weather the cotton plant is growing well and is more appealing to the insects. In other words, it does not matter what the weather is doing so far as the farmer staying with his insect control program is concerned.

Between the food requirements and the systematic insect control program it would appear the farmer will be making frequent trips through the field with his tractor. The number of trips he has to make can be reduced somewhat by combining the poison with the food. From all the accounts I have seen, it is just as good or better to include insecticides in the food solution.

Adding insecticide to the fertilizer can be done at no extra cost and the poison seems to be more effective when mixed with the plant food.

It has been found that Azodrin is effective against thrips, hoppers plant bugs and boll worms moths. But I find Pydrin more effective against the boll worm themselves. To the usual spray of 1/10 pound of Azodrin and 2 pounds of sugar in 10 gallons of water per acre, Pydrin is added at the ratio of 1 gallon per 50 acres. We always want to be looking for a better and cheaper way to control the insects, but as of right now this mixture has been found to be most efficient and effective. The reason for the mixture of Azodrin, sugar and Pydrin working is that the Azodrin in sweet water will cause the boll worm moths that drink it to die and this stops their egg laying. The Pydrin will kill what small boll worms are there. Using a small dose at regular intervals (every five days) seems to do a better job than large doses applied less often. Azodrin is also effective against red spiders and white flies when they appear later in the summer.

Maintaining the proper diet in the plants and thus controlling their growth rate is also useful in the control of insects. The bugs do not seem to be as attracted to slow growing plants as they are to fast growing plants. If the nitrogen available to the plant is either too high or too low the insects will thrive on that plant more than one with a balanced diet. At certain times the sap of the plant contains more sugar, which makes it more appealing to the insects. It has been noticed that when the plant is not fed properly and has a boron deficiency more eggs seem to hatch.

The Louisiana Cooperative Extension Service, in their June 8, 1981 report entitled "Cotton Pest Management Report No. 5", advised farmers: "Don't spray for overwintered boll weevils on cotton. There are lots of beneficial insects already in most cotton, so don't spray and kill them....you'll need the beneficial insects to keep the boll worms under control." Another report indicated, "early insect control is doomed to failure from the start". These are statements from our agricultural experts with which I take issue. This represents the traditional approach to dealing with insects. We have sacrificed the early squares to the insects to delay our spraying program in the hope of saving the later part of the crop. But as I said earlier, as much as 1/2 of the potential crop is lost this way. We put off killing the insects believing and hoping that we do more harm than good by killing them and the friendly insects along with them. This is not so. These bugs take all the crop, and all the farmer gets is what is left after they have gotten tired of eating the cotton and move on to other plants. Another problem is that by losing the early squares and bolls the plant is encouraged to put more of its energy into producing stalk instead of fruit, so the farmer loses twice, the insects get some of the bolls and because of that the plant does not produce as many bolls as it might otherwise have done. As far as killing the friendly insects is concerned, if we wait for the lady bugs to appear, the thrips have already damaged the crop. If we control the thrips early, they are not there to attract the lady bugs, and those friendly insects will go elsewhere for their food. We haven't destroyed the lady bugs, and we have saved a goodly portion of the cotton crop.

The idea appears to be wrong to let the bugs eat the first half of the crop before we begin to control them. All bug men seem to agree to let the bugs get the early squares, then poison really good to save the last crop. If you poison correctly you can have both crops. Gene Woodall, of the University of Arkansas, says, "The latter part of July and the first of August is no time to save money on poison; the insects must be under control by then". Keep the insects off the cotton early and you will be able to pick one crop in September and another in November instead of waiting for just one November crop. The way to keep the insects under control is through the use of a systematic application program as presented here.

In the early 1920's I worked out a system of insect control that produced big crops of cotton on this farm each year. The system consistently produced a much higher yield than the adjacent farms.

The first version of this device was made by me in the blacksmith shops. There was no mechanical shop on my farm. The device was used on the mule cultivators first and then on the two row tractor cultivators when tractors came into being. Later I had the device made by a local mechanic and it was made better than my crude model.

This device was used until Bezene Hexicholoride was invented. When I started using chemicals I stopped this other method because the chemical way was easier, but not as effective nor as cheap.

In a few years I sent all the devices to the junk yard to get rid of them.

About ten to fifteen years ago, more or less, I thought this old reliable system that worked well should not be forgotten and discarded. I drew up plans to get an able mechanic to produce six of these devices and I have them now put up in my shop. They have never been used. The idea will still control the bugs and produce a good crop without poison. Bugs do not get immune to this simple method. They are practical and will work and the stalk will be kept down to a reasonable height.

CHAPTER VII

WATER

Most years in the mid-South, there is enough moisture to bring a big crop to maturity. Especially if the crop is handled properly and sufficient food is furnished at the right time. Many years there will be moisture stress in the cotton for a few days and then enough rain will fall to provide sufficient moisture for the balance of the year. The farmer never knows when these periods will occur, so he must be ready to furnish moisture to his crop during that critical period. Too many farmers are too willing to rely on nature to provide the moisture, but nature doesn't always cooperate. Consequently, good cotton production depends on irrigation, even in the South.

The reason for irrigating cotton is to furnish moisture to assure a good crop even in periods of drought. My experience in 1980 serves as a good case in point. In that year this whole country was declared a disaster area due to a drought. I was called lucky because I had spent years preparing my land and learning to irrigate. My crop, in 1980, was one of the best I have ever had, while many of the farmers around me had a bad year.

Irrigation is not the entire answer to better yields, but it is necessary for top yields when the rains don't fall just right. Irrigation must be used in time before the crop stresses and it must not be used to water log the soil so that the roots cannot get air. Each boll of cotton requires a lot of water in order for it to mature. The more bolls on the plant the more water is necessary, since the plant uses water to carry the food to where it is needed.

There are many ways to furnish the moisture, and the farmer must decide what method he thinks will suit his conditions. Those methods include sprinkling, furrow irrigation, flooding, drip and the gated pipe method.

Of the methods of irrigation available, flooding is possibly the least desirable, but some few isolated farmers practice it in the western United States. The flooding system won't work in this part of the country because it will water log the land, and because of the frequency and likelihood of rain following the irrigation. Sprinkling, with its various ways and methods of getting water to the crop on time, has its followers. Sprinkling is rapidly increasing in popularity and may well surpass the furrow method in a few years, if it hasn't already, but I still prefer the furrow method for doing a timely job. If sprinkling cotton, it is best to put on 1 to 2 inches of water and move and plan to come back with more water soon to keep the plant from suffering. Read the moisture meter every day and know where the moisture is.

There is also the drip system of irrigation where a pipe is run along the rows and the water just drips all the time to furnish the necessary moisture. This is very exacting but very efficient where handled properly. These drip systems gain in acreage every year, and after they are installed the users are very successful.

Putting water in a furrow is possibly the oldest method of all. There are probably more acres irrigated by the furrow method than any other, but the sprinkler is growing in popularity. The furrow method involves digging ditches to carry the water to the point you want to run it down the field, then using syphon tubes to get the water from the ditch into the furrows.

I prefer the furrow method. The reason that syphon tubes are superior to the gated pipe method is that water can be put on the field more efficiently and with less work and more economically. Water can be put on the field with syphon tubes and not water log the soil. So often when the gated pipe system is used it will take 24 hours or longer to get the water from one end to the other. This is too long and means too much water is soaking into the soil causing water log problems. Or one finds that one end of the field is overly soaked and the other end is dried out. Syphon tubes can cover the row in 12 hours, and enough can be run down the rows at a time to get the job done correctly. The important thing about any irrigation system is to get the water to the plant on time, before it stresses, without soaking the land.

Obviously, then having water when it is needed is essential for top production. When the farmer knows he will need to irrigate, he should start getting the materials in order then, don't wait until the plant shows signs of stress. Watering must be done on time and still not soak the land. Find out what food to feed the plant and water right and a maximum crop will be made, which means 4 bales of cotton per acre and more.

The reason a lot of farmers don't profit by irrigation is that they wait too long, they wait until the 18 inch meter reading reaches zero or has been on zero for several days and the crop is hurt by the drought before they provide water. This also means that when they do water the plant has to start all over again, and then the plant gets big and thick with new growth and rots the early crop. Consequently, they claim that irrigation for cotton does not pay. Be sure to keep the cotton with enough water from the beginning and never let it run out. If this happens, the plant will never come back to make as big a crop as it would if it was never set back.

At the same time, never wet the ground to the extent of driving out all the air. Just pass the water over the top and then move on and come back in about ten days if there has been no rain. Keep the ground around the cotton plant damp at all times while is is fruiting heavily – in the months of July and August in particular. Then hope that the fall is dry so the bountiful harvest can be gathered. The roots of the plant need air and when the ground gets soaked the roots cannot get the air and this causes the plant to throw off much of the fruit it has on it. Keeping the ground moist (but not water logged) all during July and August will not hurt the plant.

I have mentioned that the farmer needs to prepare early. The time to get the irrigation system ready to operate is in mid-June so it will be operational by the end of June. Unless nature provides some unexpected rain, the irrigation will continue during the months of July and August. There is usually enough water in the soil to last until July – July and August are the months during which drought will injure the cotton. Put water on the land four or five days before the cotton gets to the point where it would suffer.

The off row needs opening up so the water can be handled properly. During long dry spells in the months of July and August, the idea of watering every other off middle every eight to ten days is right for the best production. The off rows are the spaces left between the rows during planting. (See the Chapter on Planting).

Irrigation is necessary to good cotton production because the farmer cannot rely on adequate rain always being available, and because the water used to poison insects and feed the plant does not do very much to help meet the moisture requirements of the cotton. There simply is not enough water used in spraying to provide the moisture requirements the plant needs. Ten gallons of water to the acre does not amount to much water.

Depending on the amount of rain received and the moisture meter readings, it may be necessary to begin the irrigation of the plant in late June. It is all right to water the crop when the eight inch meter reading is ten percent in late June because the plant doesn't have much fruit on it yet. The last date to water the plant will often be in mid-August, but the feeding schedule should continue for several more feedings of nitrogen. The chances are very good that no more irrigation will be needed after August.

I have noticed over the years that we frequently get a series of June rains. These rains in particular seem to cause the nitrogen to go out of the soil. This means June rains may require heavier feeding. The safest procedure to follow is to maintain a regular program of taking petiole samples and pay attention to the advice your chemist provides. A July rain tends to cause the cotton plant to use the nitrogen it has available faster than normal, so once again special attention to the feeding program may be necessary because the cotton plant needs a lot of nitrogen to produce bolls and July and August are the prime production months.

Water is needed by the plant to carry food to the leaves. If the water has ample food in it, it will not take much water to keep the plant healthy. Water and food work together in the plant.

If the plant does not have food, even though it has sufficient water, it would still cut out and shed its fruit. With our present knowledge of foliar feeding, the farmer can load the stalk up to the top with bolls provided he gets enough water to it.

When the temperature is 95 to 100 degrees, a lot of moisture leaves the soil. Even so, the cotton should still be growing and holding all squares. The cotton does not require as much moisture to sustain growth as it does to mature its fruit. The more bolls on the plant the more water and nitrogen needed to fill them out.

The method of irrigation I prefer is the furrow method. This means running the water down every other off middle when planting 2 and 1 and about every fourth middle when planting solid. I try to get the water to flow fast enough that it will move on down the row in about twelve hours and not soak the soil and thus drive the air out of the soil. Given the amount of water I pump into the ditch, I use two inch tubes, usually running three tubes to a row. The water should run out the end of the row in twelve hours. It is important to note in using syphon tubes that an increase in the size of the tube has a geometric impact on the quantity of water carried. To double the size of the tube would provide four times the amount of water. Irrigating every other off middle row will provide sufficient moisture. Use as many tubes on each row as necessary to get the water to run out the other end of the row in twelve hours.

A properly engineered ditch is important to efficient and effective irrigation. The ditch should be surveyed to provide a one tenth slope. This will provide the proper water flow. A properly engineered ditch is one which is perfectly uniform. It has no low or high spots in it and is 6 to 7 feet wide at the top. If the ditch is engineered properly, one man should be able to water 400 acres.

Plastic dams are needed to hold the water at the point on the ditch where the tubes have been placed. Sometimes the proper flow of water will necessitate putting "doors" in the dam. A door is an adjustable slack in the middle of the dam to allow a slight flow of water over or through the dam to prevent the water from over flowing the banks of the ditch.

Irrigation is not hard if the engineering is right. If the land is prepared right, the ditches are opened properly and the delivery or pumping system is functioning, the job of putting the water on the crop is neither hard nor expensive. It is not often that enough rain falls at the right time to produce a big crop. If the farmer wants top production, he must be prepared to furnish water at the needed time. It will pay handsome dividends if it is done intelligently, but if it is done in a slip shod fashion it is not of much value.

Knowing the proper time to get the water to the plant is provided by the use of moisture meters in the field. Once the 18 inch reading gets to zero the farmer has two or three days to get the water to the plant, but most fields are too large to be able to water in time. It takes ten to twelve days to irrigate 100 acres. When the ground moisture meter reads 10 percent for eight inches and 20 percent for 18 inches it is time to apply water to the plant. When the long summer days have temperatures in the upper 90's or close to 100 degrees the water will disappear in a hurry. By the use of moisture meter blocks planted in the ground after the crop is up, the farmer will be able to tell far enough in advance when to irrigate to keep the crop from suffering. Today the moisture meter will tell us when to start irrigating the crop so there is no guess work involved. It is impossible to just look at the land and tell how much moisture is in the soil.

There is one thing that is very necessary that is not stressed enough with regard to using any irrigation system, the farmer must know the moisture content of the soil to do an intelligent job. If he guesses at the soil moisture he is in for useless trouble and expense. It is necessary to have moisture blocks in the ground at regular stations in the field and read what these moisture stations say and then irrigate accordingly. When the leaves wither, most of the time that means it is too late to put water on for good results. The plant may be saved after this and new crop started, but the bolls and squares on the plant will be shed and that part of the crop lost.

The proper land slope and high rows are also important to effective irrigation. Where the land has been prepared with less than a three tenths slope the water does not run correctly and will jump from row to row. Where there is more than a three tenths slope the water runs too fast and does not get into the soil and thus into the plant. Good high rows are important to help the water run down the rows properly, to keep the water in the furrow and prevent it from spreading across the field. Without high rows the water will soak the soil too much and will not run out the end.

While the farmer's concern turns to providing water, he cannot afford to neglect his program of feeding and insect control. If the insects have not been cleaned up the plant will get into a growing stage and will abort many young squares and the internodes will grow long, from 5 to 7 inches, instead of the desired two inches or less. Nor can the farmer afford to slack off on the feeding schedule. In fact, as was pointed out earlier, the continued feeding program is essential because the water being provided to the plant is used to carry the food to the leaves, without the food, the water is wasted. So many things have to work together and on time for a good cotton crop to be produced, insect control, food and water. But the farmer should always remember, don't ever let the cotton stress for water, not even for one day, if he really wants to make as bountiful a crop as possible.

In my desire to do a good job of preparing the 40 acres used to do my work on, one of the basic laws of nature was disobeyed. The big tractors with the deep plowing tool the big ox subsoiler was used, going 2 ways to surely loosen it all up.

The powerful tractors did the job when the soil was too wet and injured every foot of the land of this 40 acres. The plant was not able to send its roots deep down in the soil to gather food and water. The cotton roots will not grow in mud and water. In all probability what happened was the subsoiler running when the land was too wet, destroyed the ability of the moisture to move up and down through the soil.

The rains that fell during the winter and spring did not go on through the ground but went down to this layer of injured land and could not get through and just stopped there.

In the month of May and June when using the soil probe to see the condition of the soil, the top 4″ or 5″ was moist but there was mud and water on down deeper than that. I wondered why? I thought at the time that this would help the plant growth, but have found out by this that it will hinder growth and development.

CHAPTER VIII
HARVEST

Surprising as it may sound, it really is much easier to make a crop than it is to gather it. The matter of getting the squares to stick on the plant and bringing them to full mature bolls of cotton is something the farmer can control and do something about, as we have discussed in the previous chapters. The farmer's ability to harvest the crop depends on good weather; having the fields in a condition that the pickers can get into them. The new pickers our technology has provided will pick the cotton if we have good weather in which to do so.

The time to pick a boll of cotton is when it has opened and dried out. Of course this cannot be done, but it should be picked as soon as possible. That is why I recommend two pickings. I like to pick the first time by September 15. This will gather the early crop which was saved by the use of the insect control program discussed in Chapter IV. The second harvest takes place in late October or early November. This harvest is the one most susceptible to the weather.

I have mentioned on several occasions that farming is a timely job, and harvesting the crop involves proper timing as well. The pickers should be checked out and repaired if necessary. It is wrong to run pickers that do not get all the cotton, because an improperly functioning picker will waste enough cotton in the field to pay for the repair. The farmer should also be sure enough modules are available so the pickers won't have to be stopped when the crop is right. Enough good pickers should be available to keep up with the crop when it opens.

One way to be sure enough pickers and modules are available is to swap out with your neighbors. If one or more farmers swap out picking then when a field is ready it can be picked. More cotton can be ginned this way. In past years my son has swapped out picking with his neighbors and it has always worked to the advantage of all. When ten pickers get in a field, the field is covered fast and most of the time it is picked on time. A module picks the cotton and stays in the field, it does not go immediately to the gin. If the cotton is put into a trailer then you take the trailer from the field to the gin. Either transportation system will help you pick it when it is ready, and as mentioned before, cotton should be picked as soon as it is open.

I believe that all farmers should have some of their crop scheduled to be picked in September and carry some of it on to October for higher production. This way the pickers are used more efficiently and the farmer assures himself of some income in case of bad weather. I am convinced that two pickings is definitely the way to do it. Just take the top eight spindles out of the picker and substitute nub spindles. If you are diligent with this you can pay your debts on the first picking and have the second picking for your pocket. However, after picking the first time you should then poison and feed the stalk again so it will continue to work for another thirty days. You will be able to defoliate around the last of October and pick the remainder of the crop then. By removing the top spindles the part of the plant continuing to grow will not be hurt. Cotton should be ready to pick six days after defoliation.

The influence of the weather, at least in northern Louisiana, is much more serious at the time of the second harvest, because it is at that time that we often have fall rains. You don't want to pick cotton when the plant is wet. If the plant has too much dew on it or a light rain has fallen, the dampness on the plant could clog the pickers.

Should the rain continue, as they did in 1974 when it rained continuously and all the cotton rotted, the idea of having two harvests will surely save the farmer from going broke. When we have excessive dampness we have the danger of the mature bolls rotting. When we have continuous rain there is little that can be done, but in years of intermittent rain we do have something that will help. We can use Isoback. This material was made by Webb Wright in Ft. Myers, Florida. When the cotton boll is just opening, the fiber is wet and it needs warm dry weather to cause it to fluff out and be a healthy boll. At this stage the mold germ attacks the boll and will not let the fiber fluff out, creating what we call a hard lock boll. Damp weather contributes to this. Isoback kills the mold germ and prevents it from growing on the fiber and causing hard lock bolls. The greatest danger we face is the cotton rotting in the field before it can be harvested. Isoback has been a great aid to cotton farmers in this regard.

If we understand the basic nature of the plant, we can learn to prepare ourselves for what is going to happen with the plant, and thus do a more timely job. The blooms that appeared on or before July 15 will be mature bolls ready to pick by September 15, or in 60 days. Knowing that, the farmer can gauge the time needed to have everything ready. That means the farmer will have some lay-by time. Lay-by means that there is no more work with the plant the farmer can do. He can only wait for the plant to open so it can be picked. The smart farmer uses the lay-by time wisely and is ready when the plant is ready.

Remember, pick the cotton as soon as you can get into the field when the bolls are open. I generally pick it twice. I have tried a third picking but have found it is rarely worth while to do so. After the last picking I go through the field with a stalk cutter, which is a separate machine used after the picker has finished, that cuts down the remaining stalks and helps get the field ready for preparation again.

Harvesting time also provides an excellent opportunity for the farmer to see how effective his soil preparations and feeding techniques have been. One can often notice how the larger plants, in terms of stalk size, do not have as many bolls on them. The impact of sloping land as opposed to level land can also be seen in the quantity and quality of the harvest. Level land tends to produce small stalks with very few bolls. The whole plant in that type of land struggled too much to produce well. In 1980 I noticed that my first picking was reduced significantly, perhaps as much as a bale of cotton per acre, because I had not begun the insect control program when I should have and had lost too many squares to the insects.

During the lay-by time the farmer may be interested in having an estimate of his crop. It takes 120 bolls per ten feet of row to make a bale of cotton per acre. The farmer should also be advised that having the bolls on the plant and getting the harvested cotton to the gin may well be two different things. Still, if he follows the advice given here, he should be able to harvest nearly all of the cotton his efforts have put on the plant.

Experience over the years has shown that there should be about three different crops in order to be assured of a profit. Cotton, wheat and corn (or milo) will divide the work over the year; one won't interfere too much with the other and will give some income. Plant the wheat in October behind the corn. This will come off in June and the land can be prepared well and set up for cotton the following year. Then when the cotton is off, hip the land high and plant corn on this land in March. This corn can be gathered in July and August and the land prepared for wheat in October. Do not try to double crop after the wheat is off, just prepare it for corn the following year and have the drainage and seed bed right. If a person has 300 acres he can plant 100 acres of corn or milo and 100 acres of wheat each year and he will prosper and not take too much labor and he will be prosperous by using his head. The wheat will take no hand labor and the land will be fallowed which is the best fertilizer possible and your crops after it will be good. The cotton land would have already been prepared. After the cotton is gathered the land need only be hipped up and planted to corn in March. Atrizine will suffice for cultivation and after the corn is gathered in August there will be ample time to drain perfectly and get the land ready for wheat to be planted in October. You will only have

100 acres to work at any given time and you will be on a sound basis.

One crop release food that causes another plant to grow so we diversify our planting so that the plant can take advantage of this.

There are three things that always pay off handsomely for a farmer and the profits will always be great.

1. Open all ditches when the ground is dry.
2. Prepare the land to plant when it is dry.
3. The third most valuable thing is to make the seed bed high, doing as much preparation as possible before the rain and wet weather sets in.

When the previous crop has been gathered, the stalks in, the land should be hipped up over the old stalks and left alone. The sleet and snow and ice will make a good seed bed for the next crop. The land does not need breaking for best production. Before planting just rehip the bed to destrop the vegetation. Drag off the plant when the ground is warmed up properly and you are on the way. Leave the bed as high as is practical so the sun can warm it up fast.

This way will let the soil be more porous and let more water in as well as let more air in, for the soil needs air for the roots to function better. When a person does this, it shows that he has learned this from experience and realizes that nature does a far better job of preparing the seed bed than he can do.

It rains in this country. A ditch that carries the water off is the most valuable property you own. A farmer who used common sense and opens his ditches just before any other job is wise and shows that he had had experience. He will always be one of the best farmers in the country as well as being one of the most properous. His crops will always be among the best in the vicinity.

Experience has shown that a three to one slope is about right with ditches and canals deep enough to lower the water table below the seed bed. When this is done the seed bed is bigger and more suitalbe for the roots to feed in.

All efficient farmers have learned this from experience. All those farmers who expect to be efficient and prosperous must practice this before they are successful.

When a man makes the decision to put these three things first, then he had taken the first steps toward being successful and prosperous.

He can then proceed to other things that will help to increase your income such as:

1. Using the right kind and amount of fertilizer.
2. Using the right plant population.
3. Using the right kind of seed.
4. Using the right kind of weed control.
5. Using the right kind of insect control.

6. Using the right kind of cotton picker.
7. Using the right kind of tractor.
8. Using the right kind of implements.
9. Handling the labor correctly.
10. Proper ginning.
11. Paying your bills on time.

By adhering to the first three things you will have something to pay your bills with. If you don't adhere to those first three things, lots of years you will have nothing to pay with, even though you planted the right seed, had good tractors and implements, had plenty of good labor, used plenty of the right fertilizer. The trouble you will have often is that you won't have much to gin and sell.

THE COTTON CROP OF 1925

(States ranked by total production)

STATE	PLANTED ACRES	HARVESTED ACRES	YIELD PER ACRE/LBS.	TOTAL BALES
Texas	18,443,000	17,336,000	115	4,163,000
Mississippi	3,644,000	3,618,000	263	1,991,000
Oklahoma	5,396,000	5,288,000	153	1,691,000
Arkansas	3,454,000	3,385,000	236	1,600,000
Alabama	3,341,000	3,314,000	195	1,352,000
Georgia	3,372,000	3,288,000	169	1,164,000
North Carolina	1,825,000	1,798,000	291	1,097,000
Louisiana	1,750,000	1,727,000	252	910,000
South Carolina	2,302,000	2,267,000	187	889.000
Tennessee	1,163,000	1,146,000	215	515,000
Missouri	530,000	511,000	280	299,000
California	174,000	172,000	339	122,000
Arizona	162,000	162,000	350	119,000
New Mexico	142,000	109,000	290	66,000
Virginia	101,000	100,000	277	58,000
Florida	114,000	112,000	182	43,000
Others*	59,000	57,000	214	26,000
TOTAL U.S.	45,972,000	44,390,000	173.5	16,105,000

*Illinois, Kansas & Kentucky

TWO HUNDRED YEARS OF U.S. COTTON

DAN P. LOGAN

COTTON FARMER

GILLIAM, LOUISIANA 71029

AVERAGE ANNUAL SPOT PRICE OF COTTON PER POUND CROP YEARS 1731-32 TO 1986-87

CROP YEAR	CENTS PER POUND	CROP YEAR	CENTS PER POUND	CROP YEAR	CENTS PER POUND
		1750-51	26.8	1770-71	17.3
1731-32	12.7	1751-52	27.9	1771-72	16.1
1732-33	14.3	1752-53	22.1	1772-73	20.0
1733-34	13.6	1753-54	22.4	1773-74	19.1
1734-35	14.3	1754-55	18.7	1774-75	23.8
1735-36	13.8	1755-56	19.7	1775-76	29.4
1736-37	15.5	1756-57	24.0	1776-77	----
1737-38	15.1	1757-58	19.2	1777-78	----
1738-39	17.8	1758-59	19.8	1778-79	----
1739-40	17.8	1759-60	18.5	1779-80	----
1740-41	16.7	1760-61	18.6	1780-81	
1741-42	13.4	1761-62	23.4	1781-82	30.5
1742-43	13.3	1762-63	25.1	1782-83	38.2
1743-44	14.0	1763-64	29.2	1783-84	32.0
1744-45	17.5	1764-65	27.6	1784-85	31.4
1745-46	21.5	1765-66	26.3	1785-86	35.9
1746-47	25.7	1766-67	27.5	1786-87	40.2
1747-48	25.8	1767-68	23.4	1787-88	38.2
1748-49	21.2	1768-69	20.2	1788-89	30.5
1749-50	24.0	1769-70	17.1	1789-90	26.3

Average Annual Spot Price of Cotton Per Pound (Cont'd)

CROP YEAR	CENTS PER POUND	CROP YEAR	CENTS PER POUND	CROP YEAR	CENTS PER POUND
1790-91	21.9	1820-21	14.9	1850-51	12.6
1791-92	30.6	1821-22	14.7	1851-52	9.3
1792-93	34.6	1822-23	11.2	1852-53	11.0
1793-94	31.6	1823-24	14.7	1853-54	11.0
1794-95	32.2	1824-25	17.9	1854-55	10.3
1795-96	31.8	1825-26	13.4	1855-56	10.3
1796-97	28.5	1826-27	10.2	1856-57	13.2
1797-98	29.4	1827-28	10.3	1857-58	12.5
1798-99	29.8	1828-29	9.9	1858-59	12.1
1799-00	29.9	1829-30	9.7	1859-60	11.3
1800-01	28.7	1830-31	10.2	1860-61	12.3
1801-02	24.0	1831-32	9.1	1861-62	27.3
1802-03	18.3	1832-33	11.4	1862-63	47.0
1803-04	18.1	1833-34	13.1	1863-64	54.2
1804-05	22.0	1834-35	16.3	1864-65	47.0
1805-06	23.5	1835-36	16.7	1865-66	31.4
1806-07	22.1	1836-37	14.4	1866-67	22.9
1807-08	17.8	1837-38	10.0	1867-68	17.5
1808-09	15.0	1838-39	13.2	1868-69	20.9
1809-10	14.8	1839-40	9.4	1869-70	20.7
1810-11	13.6	1840-41	9.9	1870-71	15.2
1811-12	9.3	1841-42	8.1	1871-72	19.5
1812-13	13.0	1842-43	7.2	1872-73	17.6
1813-14	25.5	1843-44	7.8	1873-74	15.5
1814-15	23.1	1844-45	5.9	1874-75	13.9
1815-16	28.3	1845-46	7.8	1875-76	11.5
1816-17	27.1	1846-47	10.9	1876-77	10.9
1817-18	31.0	1847-48	8.5	1877-78	10.9
1818-19	24.7	1848-49	7.2	1878-79	10.8
1819-20	16.5	1849-50	12.0	1879-80	12.1

Average Annual Spot Price of Cotton Per Pound (Cont'd)

CROP YEAR	CENTS PER POUND	CROP YEAR	CENTS PER POUND	CROP YEAR	CENTS PER POUND
1880-81	11.4	1910-11	14.9	1940-41	11.1
1881-82	12.1	1911-12	10.9	1941-42	18.3
1882-83	10.8	1912-13	12.3	1942-43	20.2
1883-84	10.9	1913-14	13.2	1943-44	20.6
1884-85	10.7	1914-15		1944-45	21.9
1885-86	9.5	1915-16	12.0	1945-46	26.0
1886-87	9.9	1916-17	19.3	1946-47	34.8
1887-88	10.2	1917-18	29.6	1947-48	34.6
1888-89	10.4	1918-19	31.0	1948-49	32.2
1889-90	11.3	1919-20	38.3	1949-50	31.8
1890-91	9.5	1920-21	17.9	1950-51	42.6
1891-92	7.7	1921-22	18.9	1951-52	39.4
1892-93	8.5	1922-23	26.2	1952-53	34.5
1893-94	7.8	1923-24	31.1	1953-54	33.5
1894-95	6.4	1924-25	24.7	1954-55	35.0
1895-96	8.1	1925-26	20.5	1955-56	35.5
1896-97	7.7	1926-27	15.2	1956-57	33.5
1897-98	6.4	1927-28	20.4	1957-58	34.4
1898-99	6.0	1928-29	19.7	1958-59	34.5
1899-00	8.4	1929-30	16.6	1959-60	31.8
1900-01	9.4	1930-31	10.4	1960-61	31.3
1901-02	8.7	1931-32	6.3	1961-62	34.8
1902-03	10.0	1932-33	7.4	1962-63	34.5
1903-04	12.8	1933-34	11.1	1963-64	34.3
1904-05	9.1	1934-35	12.4	1964-65	31.9
1905-06	11.3	1935-36	11.7	1965-66	30.7
1906-07	11.2	1936-37	12.9	1966-67	23.8
1907-08	11.5	1937-38	8.8	1967-68	30.0
1908-09	10.2	1938-39	9.0	1968-69	25.5
1909-10	14.7	1939-40	10.3	1969-70	22.2

Average Annual Spot Price of Cotton Per Pound (Cont'd)

CROP YEAR	CENTS PER POUND	CROP YEAR	CENTS PER POUND	CROP YEAR	CENTS PER POUND
1970-71	23.6	1976-77	70.9	1981-82	60.5
1971-72	31.5	1977-78	52.7	1982-83	63.1
1972-73	31.3	1978-79	61.6	1983-84	73.1
1973-74	55.7	1979-80	71.5	1984-85	60.5
1974-75	41.7			1985-86	60.0
1975-76	58.0	1980-81	83.0	1986-87	53.2

NOTES:

1. Data compiled from the files of Dan P. Logan, Sr., U.S. Dept. of Agriculture and the New York Cotton Exchange.

2. Prices are yearly average spot price from the major cotton markets. (Market year is August 1 thru July 31) Grade and staple varies over time, but is what was standard for that period.

HISTORICAL NOTES:

1. September, 1864 highest price of $1.89 per pound FOB New Orleans and then declined rapidly.

2. June 10, 1932 lowest price of $.0491 per pound FOB Memphis.

3. September 24, 1973 price was $.99 per pound FOB Memphis.

Dan Logan says "When cotton prices are high, they come down, and when they're low, they go up. The trick is to know what to do and when."

LOGAN'S IDEAS OF THE MOST VALUABLE THING ON A FARM

Even though I farm with the best knowledge in the world and have not drainage, the crop will not produce a profit.

If my land is as rich as the Nile Valley and is not drained it will produce very little. I may have the best equipment with the best tractors money can buy, but with no drainage I am destined to go broke.

I may have many degrees from outstanding agricultural colleges, and do not know or practice drainage, my degrees will be of no value. I may get up early and work hard with plenty of finances but do not know how to drain or practice drainage, my knowledge and finances will amount to nothing.

Drainage makes it possible for the crop to grow early in the Spring so the plants can withstand cool, wet weather. Those who practice it will be rewarded with good crops. The crops grow well in well-drained land. Drainage is never wrong and never lets the farmer down in times of stress.

The farmer who works poorly drained soil will pass away, for it will not pay expenses over the years. Most of the time in the delta the difference between a good farmer and a poor one is that one farm is well drained and one is not.

The farmer who practices drainage and looks after his crop will prosper and thrive, and those who do not will stumble and fall and go by the wayside. The farmer who does not drain blames the weather and many other conditions that he will say is beyond his control. When in reality the wet soil was not allowed to produce its best because of its wetness. The blame is with the farmer who hasn't the foresight or the skill to get the excess water off.

The best drained land always gives the best crops. Drainage is essential to profit. Drainage is essential and comes first in all our efforts.

Therefore, get drainage first, then all these other things can be added and a profit on top of this.

Dan P. Logan, Sr.

Cotton Farmer
Gilliam, LA

TILLING THE GOOD EARTH